Eco-pragmatism

Eco-pragmatism

Making Sensible
Environmental
Decisions
in an Uncertain
World

DANIEL A. FARBER

The University of Chicago Press
Chicago and London

DANIEL A. FARBER is Henry J. Fletcher Professor of Law at the
University of Minnesota. He is coauthor of *Law and Public Choice*
(1991), also published by the University of Chicago Press.

The University of Chicago Press, Chicago 60637
The University of Chicago Press, Ltd., London
© 1999 by The University of Chicago
All rights reserved. Published 1999
08 07 06 05 04 03 02 01 00 99   1 2 3 4 5

ISBN: 0-226-23806-7 (cloth)

Library of Congress Cataloging-in-Publication Data

Farber, Daniel A., 1950–
    Eco-pragmatism : making sensible environmental decisions in an
uncertain world / Daniel A. Farber.
        p.   cm.
    Includes bibliographical references and index.
    ISBN 0-226-23806-7 (alk. paper)
    1. Environmental policy.   I. Title.
GE170.F37   1999
363.7'056—dc21   98-40507
                                                            CIP

♾ The paper used in this publication meets the minimum requirements
of the American National Standard for Information Sciences—
Permanence of Paper for Printed Library Materials, ANSI Z39.48-
1992.

TO NORA

[A]ll life is an experiment. Every year if not every day we have to wager our salvation on some prophecy based on imperfect knowledge.
Oliver Wendell Holmes

One planet, one experiment.
Edward O. Wilson

CONTENTS

ACKNOWLEDGMENTS

This book would not have been possible without the help and support of many people, not only during its writing, but also beforehand. I have struggled with these issues since I began teaching environmental law, and I owe a great debt to a host of students and colleagues for discussions during that period and for help in my earlier efforts to address some of the issues in law review articles. In particular, I should mention Paul Hammersbaugh's collaboration in the study of discounting that led to chapter 5. With regard to the writing of the book itself, there are many to thank. Andy Hall and Shannon Fisk contributed editorial and research assistance, and Laurie Newbauer carried a heavy burden of word processing. I also owe thanks to participants at faculty workshops at Georgetown, Harvard, Illinois, Minnesota, and Vanderbilt, where I presented chapters of the book, and to the following individuals for comments on earlier drafts: Jim Chen, Don Elliott, Eric Freyfogle, Clay Gillette, Richard Lazarus, Joshua Sarnoff, Mark Seidenfeld, and Gerald Torres. An anonymous reviewer for the Press offered comments that proved helpful in shaping the final product, and my editor, John Tryneski, offered moral support throughout.

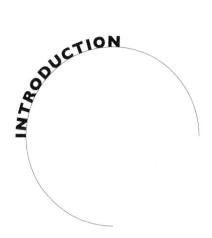

INTRODUCTION

We have made a profound national commitment to environmental protection. Environmental law, once a scattering of obscure statutes regulating federal lands, has become a jungle of complex regulations and court decisions. We are only now beginning to chart this new field. Environmental regulation requires us to make decisions under the most difficult circumstances. Seemingly incomparable values like human life and economic cost are at stake. The trade-offs are obscured by scientific uncertainty and involve effects spanning decades, if not centuries. We now sense the importance of environmental values, but we are still learning how best to make hard environmental decisions.

Evidence abounds of our commitment to environmental goals. In 1972, an estimated twenty million Americans participated in Earth Day. Thousands of colleges, high schools, and elementary schools and communities took part. Some twenty years later the reverberations were still being felt as millions celebrated the anniversary of Earth Day.[1] Students of public opinion now call environmentalism a "consensual" value in American so-

---

1. Robert Cahn and Patricia Cahn, "Did Earth Day Change the World?" *Env't*, Sept. 1990, at 16, 18–19, 37; Council on Environmental Quality, *Twentieth Annual Report* 4–5 (1990).

ciety.[2] In 1989, 80 percent agreed that "[p]rotecting the environment is so important that regulations and standards cannot be too high, and continuing environmental improvements must be made regardless of cost."[3] As a high official in the Bush administration once put it:

> Increasingly, we are all environmentalists. The President is an environmentalist. Republicans and Democrats are environmentalist. Jane Fonda and the National Association of Manufacturers, Magic Johnson and Danny Devito, Candace Bergen and The Golden Girls, Bugs Bunny and the cast of Cheers are all environmentalist.[4]

Environmentalist attitudes are found in publications that cater to a broad range of groups, including farm and automotive magazines.[5] When the Republican leadership launched a major attack on environmental regulation after their triumph in the 1994 congressional elections, they were soon dismayed by polls show-

2. Riley Dunlap, "Public Opinion and Environmental Policy," in *Environmental Politics and Policy: Theories and Evidence* 87, 98, 123–124 (James Lester ed., 1989).

3. Roberto Suro, "Grass-Roots Groups Show Power Battling Pollution Close to Home," *N.Y. Times,* July 2, 1989, at 1. See also Richard L. Berke, "Oratory of Environmentalism Becomes the Sound of Politics," *N.Y. Times,* Apr. 17, 1990, at 1 ("The environment . . . has reached the forefront of American politics, with candidates for one political office after another proclaiming themselves environmentalist."). For more recent polling data, see Sharon Buccino and Gregory Wetstone, "Environmental Policy Battles in the Congressional Budget Process: The 104th Congress: Back-Door Assault," 27 *Envtl. L. Rep.* 10113, 10116 n.50 (1997).

4. Mark Sagoff, "The Great Environmental Awakening," *Am. Prospect,* Spring 1992, at 39, 39.

5. Mark Sagoff, "Settling America or The Concept of Place in Environmental Ethics," 12 *J. Energy, Nat. Resources & Envtl. L.* 349, 414–416 (Summer 1992). Furthermore, environmentalism is surprisingly strong even in underdeveloped countries. For example, 29 percent of those surveyed in Mexico volunteered that environmental problems were among the most serious facing that country, while 45 percent of Nigerians rated their country's environmental problems as "very serious." Riley Dunlap, George Gallup, and Alec Gallup, "Of Global Concern: Results of the Health of the Planet Survey," *Env't,* Nov. 1993, at 6, 9, 10.

ing that "even a majority of Republicans say they do not trust their own party on environmental matters."[6]

Thus, environmentalists can rightly claim broad public support for environmental values. The federal laws implementing these values have achieved much of which we can be proud. Even its critics concede that the "present system has done much to improve environmental quality."[7]

Our society has hopefully reached the point where the legitimacy of those values can be taken as a given. The harder question is the priority of those values. Do they, as some insist, trump other interests such as economic needs? Or is some kind of balancing required? And if we *are* going to engage in trade-offs, how do we measure the values at stake?

As important as environmental values are, at some point it would become absurd to pursue them at all cost. In a recent book, Justice Stephen Breyer contends that some environmental rules require expenditures that defy common sense in the name of eliminating minuscule, if not imaginary, risks. One of his examples is a Superfund case in which the government demanded that $9 million be spent to remove a trace of toxic chemicals from soil. Even without this final cleanup, according to Justice Breyer, the site would have been clean enough for children to safely eat small amounts of dirt about once a week. The final cleanup would have made it safe to eat dirt about four days a week—but there weren't any children on the site in the first place because it was in the middle of a swamp![8]

Justice Breyer also discusses a 1990 study of asbestos removal from schools. According to the study, the annual risk to children from undisturbed asbestos in schools is probably less than one in ten million. Eliminating this risk will cost somewhere between

6. "Republican Survey Finds Deep Concern over GOP Environment Agenda," *Inside EPA Weekly Rep.*, Jan. 26, 1996, at 1, 8. See also Richard L. Berke, "In a Reversal, G.O.P. Courts the 'Greens,'" *N.Y. Times*, July 2, 1997, at A1.

7. Terry Davies, "Critically Evaluating America's Pollution Control System," 130 *Resources* 17, 17 (Winter 1998).

8. Stephen Breyer, *Breaking the Vicious Circle: Toward Effective Risk Regulation* 11–12 (1993).

$50 to $150 billion (not to mention the possible health hazard to the removal workers). The trade-off, if these estimates are accurate, would be something like $250 million per life saved. Justice Breyer questions whether this is a remotely sensible decision. If we translate the same trade-off to the arena of auto safety, he points out, we would pay about $50,000 more per new car to cut the present automobile death rate by 5 percent. Similarly, he says, one Environmental Protection Agency (EPA) regulation required spending hundreds of millions of dollars to eliminate certain uses of asbestos that turned out to be a smaller public health concern than the occasional death caused by swallowing toothpicks.[9] Another critic estimates that the federal Superfund program spends $4 billion for every life saved.[10]

Some of these figures are open to debate. So is the implication that similar disparities between risks and remedies are common in environmental law.[11] Environmentalists may well question the assertion that current environmental laws now commonly require such one-sided trade-offs. But critics like Breyer are on solid ground in suggesting that such trade-offs would be outside the bounds of common sense. At some point, we do have to admit that the cost of protecting the environment exceeds the benefit. The difficult question is how to go about drawing the line.

There seems to be no escape from making hard choices about when to sacrifice environmental values for other pressing concerns. In this book, I will try to work through some of the core issues in making hard environmental decisions. For the sake of concreteness, I will ground much of the discussion on one of the leading cases in environmental law, *Reserve Mining Co. v. United States*.[12] In *Reserve Mining*, the court had to decide whether to require the company to spend almost a quarter of a billion dollars to combat a health risk that might or might not actually

9. Id. at 12–14.
10. Kip Viscusi, "Regulating the Regulators," 63 *U. Chi. L. Rev.* 1423, 1436 (1996).
11. See Lisa Heinzerling, "Regulatory Costs of Mythic Proportions," 107 *Yale L.J.* 1981 (1998) (debunking the most frequently cited study of excess regulatory costs).
12. 514 F.2d 492 (8th Cir. 1975).

exist. After exploring the facts of the case and some general questions about the status of environmental values, I will then consider how existing approaches to environmental policy would address the court's dilemma. This discussion, in turn, serves as the springboard for analyzing some pervasive questions in environmental law.

Hard Choices and Contesting Dogmas

As *Reserve Mining* illustrates, environmental law often involves long-run risks only recently discovered by science and still subject to great uncertainty. Our knowledge of environmental problems is in many ways still limited. As Chris Stone of the University of Southern California Law School, a staunch environmentalist, explains:

> We are only beginning to learn how the world works. Our ignorance is not only about the dynamics of globe-spanning climate and current. Scientists have only started to inventory the world's forests and monitor the thickness of the ice caps. . . . As for biodiversity, we do not know how many species there are to imperil.[13]

Stone does not regard this state of affairs as an excuse for complacency or inaction. Nor do I. But all the other problems of environmental policy are amplified by our ignorance of the relevant facts. It would be hard enough to draw the line between appropriate and excessive environmental regulation under any conditions. But essentially we are trying to draw that line in a dark room, illuminated only by intermittent flickers of light.

Scientific uncertainty is only one of the knotty problems that plague environmental law. Among the toughest questions are these:

• How do we decide whether an environmental rule is worthwhile? On the one hand, we have a benefit such as preventing future cases of cancer or saving an endangered species. On the other hand, we have a financial burden, which may translate into lost jobs, lower productivity, or a lower standard of living. How can we de-

13. Christopher Stone, *The Gnat Is Older than Man: Global Environment and Human Agenda* 24 (1993).

cide if the benefit is large enough to justify the cost, when the values at stake seem so different?

- What should be our baseline in making environmental decisions? Should we begin with a presumption in favor of protecting the environment or with the burden of proof on advocates of regulation? Or should we take a "neutral" stance, with no general preference for or against environmental preservation?
- Unlike many ordinary decisions, the repercussions of environmental issues play out over decades or generations. This creates another "apples and oranges" problem: How do we decide how much to spend *today* in order to obtain an environmental benefit twenty or a hundred years in the *future?* How much should people be expected to sacrifice today for a better environment tomorrow?
- Our knowledge of environmental issues is changing rapidly. When should we wait for more information before taking action? And when we decide not to wait, how do we pick an appropriate response to a problem whose outlines are only dimly glimpsed? How do we avoid getting locked into outmoded legislative frameworks?

Some people seem to think these questions have simple answers. One approach that promises such answers is cost-benefit analysis. To perform a cost-benefit analysis of a ban on asbestos, we would have to begin by determining the current level of risk posed by asbestos. This stage is called risk assessment. Given our current level of scientific uncertainty, these numbers are often debatable. Having determined the current level of risk, we would need to determine the number of lives that will be saved, and then we would need to convert this benefit into dollars so it can be compared with the cost. This would require us to assign a monetary value to the human lives saved by the regulation. Because the benefits (in the form of lives saved) are realized in the future, we would need to "discount" those benefits to present-day dollars. Having put both the costs and the benefits into a common metric, we would only need to compare the two figures to determine if the regulation produces benefits in excess of its costs.[14]

---

14. For a thoughtful argument in favor of this approach, see Susan Rose-Ackerman, *Rethinking the Progressive Agenda: The Reform of the American Regulatory State* 33–37 (1992).

Cost-benefit analysis goes beyond the simple notion that we should try to obtain a given level of environmental quality at the lowest possible expense, a goal that is now embodied in federal statutes.[15] Instead, in cost-benefit analysis, cost becomes a factor in setting our goal for environmental quality. Rather than looking for the cheapest possible way to clean up a hazardous waste site, we seek the economically optimum level of safety.[16]

This approach received a strong boost in 1981, when President Ronald Reagan issued an order requiring all government agencies to base their decisions on cost-benefit analysis except when prohibited by statute from doing so. As modified by later presidents, this mandate remains in place today. More recently, quantitative risk assessment and cost-benefit analysis were major planks in the Contract with America. One proposal for "regulatory reform" would have overridden public health mandates in favor of cost-benefit analysis of all new regulations. Another proposal would have required agencies to redo all of their past regulations in light of the new mandate.[17] The result would probably have been to hamstring the EPA.

At the other end of the spectrum from the cost-benefit analysts are those who believe that public health and environmental quality are paramount, much like constitutional rights such as free speech. Advocates of this perspective can point to a bevy of federal environmental statutes in which cost plays a secondary role. They argue that public values such as the environment cannot be captured by a cost-benefit analysis. They view economic efficiency, the value underlying cost-benefit analysis, as a mere amalgam of consumer desires, not to be considered on the same scale as moral values like protecting human life or maintaining biodiversity. In their view, private markets, in which monetary

15. Unfunded Mandates Reform Act of 1995, § 201, 2 U.S.C. § 1531 (1995).

16. Case studies of the EPA's use of cost-benefit analysis can be found in *Economic Analyses at EPA: Assessing Regulatory Impact* (Richard Morgenstern, ed., 1997).

17. See Jeff Gimpel, "The Risk Assessment and Cost Benefit Act of 1995: Regulatory Reform and the Legislation of Science," 23 *J. Legis.* 61, 79–84 (1997); Robert L. Glicksman, "Regulatory Reform and (Breach of) the Contract with America," *Kan. J.L. & Pub. Pol'y*, Winter 1996, at 1, 16–17; Cass Sunstein, "Congress, Constitutional Moments, and the Cost-Benefit State," 48 *Stan. L. Rev.* 247, 269–282 (1996).

values are assigned to costs and benefits, provide little guidance in making public decisions.

Advocates of these conflicting viewpoints have waged a bitter battle at the expense of many trees.[18] On a practical level, the economically oriented, like Justice Breyer, support some forms of environmental regulation, but argue that current environmental law has often squandered billions of dollars to little effect. Environmentalists respond that risk assessment is "not ready for prime time" and that cost-benefit analysis is a recipe for "paralysis by analysis."[19]

The battle is not limited to disputes about the practicalities of environmental regulation. It also involves sharp philosophical disputes. Cost-benefit advocates consider the environmentalists to be single-minded zealots. In reviewing Vice President Al Gore's book, *Earth in the Balance,* a leading environmental economist criticizes "[v]isionary approaches" like Gore's as "almost of necessity religious in nature." He fears that environmentalists, "like the Marxists before them," will "demonstrate little tolerance for opposing views."[20] Representing the other camp, a philosopher argues that we should not blame environmentalists who ignore the potential benefits of exploiting nature. These environmentalists, she says, are like pet owners who "disregard the opportunity costs of not eating their pets or not selling them for laboratory experiments." Disregard for such options, she says, "represents not an irrational neglect of opportunities for gain, but a contempt, even a revulsion, for the gains that could be achieved by betraying our pets" or by pillaging nature.[21]

These quotations, I should stress, are from well-respected, thoughtful scholars, not political activists or ideologues. The sharpness of this language gives a good indication of the strength of their disagreement. Of course, not all of the rhetoric is this

18. For a recent overview of the debate, see David Driesen, "The Societal Cost of Environmental Regulation: Beyond Cost-Benefit Analysis," 24 *Ecology L.Q.* 545 (1997).

19. Thomas McGarity and Sidney Shapiro, "OSHA's Critics and Regulatory Reform," 31 *Wake Forest L. Rev.* 587, 617, 626 (1996).

20. Robert Hahn, "Toward a New Environmental Paradigm," 102 *Yale L.J.* 1719, 1754 (1993).

21. Elizabeth Anderson, *Value in Ethics and Economics* 208 (1993).

heated; there are moderate voices as well. But the willingness of respected scholars to resort to extreme language is an indicator of the depth of the disagreement.

Despite their conflicts, advocates of both approaches share a belief that environmental policy can be based on a single over-riding value, whether that value is economic or environmental. Both sides seek rigid methods of making environmental deci-sions, in which the correct answer is guaranteed if only decision makers follow the right recipe. This vision of legal decision mak-ing has come under increasing attack in recent years. So long as we view environmental policy as a dichotomous choice between two competing holistic theories, we will make little headway in dealing sensibly with environmental issues.

Pragmatism and Principle in Environmental Law

In this book, I argue for a pragmatic approach to environmental problems, in which economic analysis is useful, but not control-ling. Critics of cost-benefit analysis are right that economic efficiency is an inadequate basis for environmental policy. In-deed, the "state of the art" of cost-benefit analysis would limit its ability to generate firm answers to environmental questions even if we did want to make it our sole basis for decision making. But the critics are wrong to build a wall between economics and ethics. In practice, the cost-benefit analyst needs to make nu-merous technical decisions that turn out to also involve ethical issues. Moreover, many economic insights turn out to be rele-vant to a broader policy analysis. Properly understood, then, the dichotomy between economics and value judgments turns out to be a false one.

The approach that I take in this book is part of a broader movement in legal scholarship, which is sometimes called prac-tical reasoning or legal pragmatism.[22] Legal pragmatists are, in

22. See Daniel Farber, "The Inevitability of Practical Reason: Statutes, For-malism, and the Rule of Law," 45 *Vand. L. Rev.* 533 (1992); Symposium, "The Renaissance of Pragmatism in American Legal Thought," 63 *S. Cal. L. Rev.* 1569 (1992). For related work in the broader stream of philosophy, see, e.g., Richard Rorty, *Consequences of Pragmatism* (1982); Hilary Putnam, *Realism with a Human Face* (1990).

part, reacting against the increased obsession of some other legal scholars with grand theories such as economic reductionism. A convincing analysis should be like a web, drawing on the coherence of many sources, rather than a tower, built on a single unified foundation. Intelligent analysis requires the use of theories, but as tools, not as ends in themselves. Environmental decisions involve a complex network of scientific, economic, and normative judgments. It is unlikely that we can construct a structure in which all of these considerations will point to a single conclusion. We can have better hopes of building an interlocking web of arguments that will support a decision based on diverse, overlapping considerations.

Being pragmatic does not mean the rejection of rules or principles in favor of ad hoc decision making or raw intuition. Rather, it means a rejection of the view that rules, in and of themselves, dictate outcomes. Thus, we shouldn't expect some mechanical technique to give cut-and-dried answers to hard policy questions. Hard policy decisions can't be programmed into a spreadsheet. To the extent that cost-benefit analysts purport to provide such techniques, they are doomed by their inability to capture the richness of actual policy decisions.

Yet this does not mean that we should rely merely on intuition. There is no escape from the need for deep involvement with all the details and perplexities of each specific environmental problem. But we also need an analytic framework to help structure the process of making environmental decisions. Intuition is often an unhelpful guide because environmental law concerns issues outside of our normal, everyday experience. (Answer quickly: Just how much would it be worth spending to retard the greenhouse effect by a century? A billion dollars? A hundred billion? A trillion?) Ad hoc decision making also makes it difficult to achieve even rough consistency between different decisions. Yet we need some kind of uniformity among public health decisions relating to various kinds of toxic chemicals. We also need some overall policy about issues such as preservation of various endangered species, rather than a patchwork of ad hoc decisions. Ad hoc decision making poses the risk that policy will shift radically after every election, leaving us unable to maintain a coherent policy over time.

For all these reasons, we need mediating principles to guide decision making. Rather than rigid rules or mechanical techniques, we need a framework that leaves us open to the unique attributes of each case, without losing track of our more general normative commitments.

The Road Ahead

Because my approach grows out of our society's current practice, I will not offer radical proposals for transforming the system. Neither do I believe that the present system is ideal. My argument is that the current system embodies some key social values and that understanding those values more clearly would help improve their implementation. But this does not mean that hard policy issues will magically become simple. One of the points of pragmatism is that there is no escape from the need to wrestle seriously with the particulars of a given problem. Neither pragmatism nor any other approach can offer ready answers to hard problems like biodiversity and global warming, and I will not attempt to do so in this book. My goal, more modestly, is to help us find better ways of thinking about such problems, in part by showing that the current regulatory system is more coherent than it sometimes appears. In short, although I will not argue for specific policies, I will attempt to clarify the analytical framework for considering these decisions.

In this spirit, I will offer several suggestions about how to deal with some of the puzzles raised by environmental policy. For example, I will argue that our society is saving too little for the future and that we ought to compensate by "leaning into the wind" when making environmental decisions. When using the economic technique of discounting to convert future costs into their present equivalents, we should use low discount rates in order to maintain a long-range perspective. When time spans exceed a few decades, we need to use other techniques to protect future generations. I will also argue that we should use an environmentalist baseline in regulating pollution, tempered by the use of cost-benefit analysis as a test of reasonableness. This baseline would make explicit the predominant values underlying much of our current regulatory system. In interpreting statutes, I suggest,

courts should follow a "green" canon of interpretation, construing ambiguous statutes in favor of as much environmental protection as is reasonably feasible.

Although I mean these specific suggestions to be taken seriously on their merits, they also serve to illustrate a style of thinking about environmental problems. Even readers who remain
unconvinced by my conclusions, I hope, will see merit in pragmatism as a way of analyzing hard environmental questions.

In approaching specific environmental issues, we must keep
in mind our existing commitment to environmental protection.
Rather than approaching each environmental problem afresh,
as if we had never seen one before, we need to adopt a baseline rule of eliminating environmental risks as much as feasible.
Only when the costs are grossly disproportionate to the benefits
should we abandon this baseline. We also need to keep firmly in
mind the limits of our knowledge of environmental problems.
Environmental law operates on the frontiers of scientific knowledge, and today's best estimates will be soon be replaced. When
the existence of environmental hazards is still unclear, we should
take reasonable precautions, but we should also establish systems for improving regulatory techniques as further information
becomes available. Thus, our regulations need to be environmentalist, yet flexible. When we learn that a regulatory scheme
is outmoded, the EPA should have authority to deregulate, just
as it needs residual authority to obtain interim remedies against
newly discovered environmental hazards until Congress can act.

The ultimate challenge for environmental law is social sustainability. It will do little good to save the planet today, only
to lose it tomorrow. Thus, we need an approach that not only
embodies our firm commitment to the environment, but also
recognizes competing goals and the need to keep up with changing scientific knowledge. Otherwise, we will have a regulatory
structure that is too draconian for us to live with in the long run.
Among the components of the global eco-system are the clever,
idealistic, aggressive, acquisitive creatures known as *homo sapi-
ens*. Environmental law must create a hospitable environment
for them as well as for other organisms. Environmental law must
be pluralistic and flexible if it is to endure. Eco-pragmatism is a

rough and ready approach to environmental policy, perhaps lacking in elegance, but durable enough for hard wear.

The need to make environmental law "sustainable" is a theme that runs through much of the book. It helps drive arguments on a wide range of topics. For instance, chapter 2 argues that we should reject the premise that economic interests are mere "preferences," entitled to little or no consideration compared with environmental values. Underlying the argument, in part, is a concern about sustainability. Given the nature of human behavior in modern societies, it is unrealistic to expect environmental programs based on such an austere premise to endure long. In chapter 4, for similar reasons, I argue that we should be prepared to modify environmental regulations whenever their costs are grossly disproportionate to any possible benefits. Chapter 5 discusses the extent to which current generations can realistically be expected to make sacrifices on behalf of distant descendants, and among other topics, chapter 6 considers how we can prevent outmoded regulations from eroding the overall credibility of environmental law. In taking these positions, my goal is not to undermine environmental values, but to implement them in a way that we can expect to endure, as opposed to heroic efforts that are likely to fade after a few years. Environmental protection is a marathon, not a sprint.

Some of the issues I will discuss are formidably conceptual. I will alternate discussion of these abstract issues with discussions of concrete issues in environmental regulations. Chapters 1 and 3 explore the *Reserve Mining* case and its implications, and chapter 6 considers specific regulatory techniques. Chapters 2, 4, and 5 discuss more fundamental questions: the ethical status of the goal of economic efficiency, the need for an environmentalist baseline, and the nature of our obligations to the future. These issues arise naturally out of the *Reserve Mining* case study, but inevitably move to a more abstract plane. Hopefully, periodic returns to more concrete problems will offer some relief from excessive conceptualism.

Environmental law raises many kinds of issues. In this book, I will focus on those issues that are most distinctive and characteristic of the field, particularly those involving scientific uncertainty, conflicting economic and noneconomic values, and long time-horizons. In so narrowing the focus, I am putting to the side some other important questions. Even if environmental law makes society as a whole better off, there will be winners and losers, and these distributional effects are often pivotal politically. The distributional issues loom especially large in the international arena, given the huge wealth disparities between the developed northern hemisphere and the less developed southern hemisphere. Moreover, even after we identify the correct policy outcomes, designing the right kinds of institutions to implement those outcomes is a severe challenge. Like the distributional issues, questions of institutional design loom especially large in the international arena, where the institutional apparatus is still often rudimentary. Ultimately, these institutional and distributional issues must be resolved if environmental protection is to take root on a permanent basis. These issues receive only occasional attention in this book, not because of any doubts about their importance, but only because of the necessity of keeping the scope of the book manageable.

Even with the focus thus narrowed, the subject poses great challenges. Environmental policy may be a uniquely challenging area of scholarship. It involves at least four different fields: philosophy, economics, science, and law. Mastering any one of those fields is no easy matter; to master all four would be impossible. Yet we cannot intelligently discuss the issues without making the effort. As we will see, the field has also attracted an unusually high number of top-flight scholars from all these disciplines— who, unfortunately, disagree vehemently with each other. Even putting aside differences in personal values and worldviews, no scholar can hope to have the final word on how to approach environmental issues. But at least we can hope to advance the conversation.

CHAPTER ONE

A Case of Uncertainty

We learn a lot about moral reasoning from stories about how others have made hard moral decisions.[1] Similarly, we can learn something about how to grapple with environmental decisions by observing the efforts of others. Some abstract conflicts about cost-benefit analysis or risk assessment take on a different guise when we see the difficulties of applying these approaches. In short, environmental policy needs to be grounded in the concrete realities of environmental regulation.

One of the lessons of pragmatism is the usefulness of combining concrete examples with more abstract forms of reasoning.[2] Consideration of concrete situations serves several purposes. It provides a method of testing and refining ethical principles. For instance, according to John Rawls, the best moral reasoning involves a movement between general theories and intuitions about specific issues. The general theories are tested for their fit with specific intuitions, which are themselves subject to modification if they cannot be made part of some coherent theory.[3]

1. Martha Nussbaum, in *Love's Knowledge: Essays on Philosophy and Literature* (1990), develops this thesis at length, with specific literary examples.

2. As Frank Michelman has put it, practical reason "seems always to involve a combination of something general with something specific," so that "[j]udgment mediates between the general standard and the particular case." Michelman, "Foreword: Traces of Self-Government," 100 *Harv. L. Rev.* 4, 28–29 (1986).

3. John Rawls, *A Theory of Justice* 20–21, 48–52 (1971).

Assertions like "human life is priceless and can never be sacrificed for merely economic goals" may seem less compelling after examining some concrete examples. Concrete situations—particularly those in which people are making ethical judgments—may also provide object lessons in how to make hard decisions.

It behooves us, then, to begin with a specific case study. The one I have chosen has the advantage of being relatively simple, but quite important in the evolution of environmental law. *Reserve Mining Company v. United States*[4] was the first major judicial confrontation with environmental risk. It raised troubling issues regarding scientific uncertainty, the difficulty of balancing cost against public health, and the long-term nature of environmental harms.

Reserve Mining remains a leading case on the subject of risk regulation even today. Its dramatic facts make it an ideal vehicle for thinking about environmental risks and how they should be controlled. The court was confronted with a massive discharge of asbestos, a known carcinogen, into Lake Superior. But almost all the medical evidence related to the danger of airborne asbestos. Whether there was any health risk from asbestos in drinking water was unclear. Thus, the court was faced with a massive gamble at unknown odds. At stake were hundreds of millions of dollars versus the safety of the people of Duluth, Minnesota. We could hardly invent a more arresting example of the quandary posed by environmental risks.

Background

Northern Minnesota, where the case arose, has a long and checkered history as iron mining country. Most of the high-grade deposits have long been exhausted. Reserve Mining, a joint venture of Armco and Republic Steel, was established to mine taconite, a low-grade iron ore found on the north shore of Lake Superior.[5] Taconite requires extensive processing before it can be used. Basically, the process involves smashing the taconite into small pieces, separating those with the most iron, breaking

---

4. 514 F.2d 492 (8th Cir. 1975) (en banc).
5. *United States v. Reserve Mining Co.*, 380 F. Supp. 11, 22 (D. Minn. 1974).

up those pieces, once again separating the chunks with the most iron, and so forth, through five rounds of purification. At the end, the process produces small green pellets with a high iron content. It also produces immense amounts of "tailings" or leftover bits of rock. The company disposed of the tailings by letting them flow through troughs into Lake Superior.[6]

Reserve was ultimately discharging twenty-five million tons of tailings a year into the lake. According to Reserve's own estimates, it had dumped tailings over more than a thousand square miles of Lake Superior. The manufacturing process used over two million tons of water per day, which was recycled back into the lake.[7]

Reserve Mining was not completely insensitive to environmental matters. The original construction plan called for blasting rock from a small island to use for breakwaters. A botany professor in Duluth protested because she had found rare ferns and flowering plants on the island, including the closed bluebell, not known to grow anywhere else. Reserve decided to leave the island intact, and "[a]t last report, the closed bluebells, the ferns, and a herring gull colony were all doing nicely."[8]

When Reserve began operations, it was welcomed with open arms by local communities, state politicians, and the media. The area was severely depressed, with chronic high unemployment. Reserve was an economic savior for a region that badly needed help. The enrichment process for taconite was the result of years of patient research by a public-spirited University of Minnesota professor.[9] The gratitude of local residents for the boons of taconite mining is reflected in the decision of the local Catholic Church to have its altar and baptismal font made out of taconite.

When the plant was built, Reserve obtained a permit to dispose of its tailings in the lake. Experts testified that because of their weight, the sandy tailings would flow along the bottom of the lake without mixing with the water. An aquatic biologist tes-

6. Id. at 30–31. Notably, other taconite processors disposed of their tailings on land. Id. at 64.

7. Id. at 31, 36, 38.

8. E. W. Davis, *Pioneering with Taconite* 171, 176 (1964).

9. Frank D. Schaumberg, *Judgment Reserved: A Landmark Environmental Case* 34–38 (1976).

tified that the tailings would have no harmful effect on fish. Duluth officials did wonder whether there might be any conceivable effect on the city's drinking water supply. There was also opposition from a couple of union locals and a Duluth sportsmen's club. But government agencies such as U.S. Fish and Wildlife and citizen groups such as the Izaak Walton League favored the permit. The state commissioner found the project to be in the public interest and issued the permit in late 1947. The plant sent out its first shipment of ore in early 1956.[10]

Although the corporate officers were no doubt pleased to be serving the public interest, their companies also found the venture profitable. From 1956 to 1973, Reserve returned a total of $240 million in profits after taxes. In 1973 alone, it returned a profit of about $20 million. The joint venture was highly leveraged, at least by the standards of the time, so this translated into over a 50 percent return on equity. This was well above Armco and Republic's required rate of return for investment projects. In short, the two companies had seemingly done well by doing good.[11]

The 1960s saw a great upsurge in environmentalism. By the end of that decade, massive dumping like Reserve's was no longer easily tolerated. An Interior Department report concluded that Reserve was a serious source of pollution.[12] The secretary of the interior began the cumbersome process of federal regulation under the pre-1972 federal water pollution statute.[13]

10. Davis, *supra* note 8, at 131–134, 179.

11. 380 F. Supp. at 59–61.

12. Schaumberg, *supra* note 9, at 69–72 (detailing the conclusions of the Interior Department report).

13. Enforcement under the pre-1972 Federal Water Pollution Control Act (FWPCA) could be sought under two different proceedings. The original enforcement mechanism was a conference procedure established under the 1948 law. The 1965 amendments created a second mechanism designed to provide federal enforcement of state water pollution control standards. Under this approach, abatement actions could be brought following a 180-day notice period. 33 U.S.C. § 1160 (1970). If, however, the pollution did not create interstate "endangerment," an action could be brought only with the written consent of the governor of the state where the discharge was occurring. Id. Besides greatly expanding the federal administrator's regulatory powers, the 1972 amendments streamlined enforcement mechanisms.

As Reserve's activities received continued attention, troublesome questions arose about the effect of the discharges on the lake. Total discharges by the plant were thirty times the annual sediment from all the lake's U.S. tributaries combined. Tailings suspended in the water created a plume of "green water" for a distance of at least eighteen miles. State water-quality limits for iron, lead, and copper were violated, along with recommended standards for zinc and cadmium. Because of the lake's natural "extreme clarity and cold temperature," increased cloudiness might pose a threat to the ecology, particularly to the small edge of the lake shallow enough for fish to spawn.

On the other hand, there was little direct evidence of harm to fish or other aquatic life.[14] Superior is valued in large part for aesthetic reasons, but this was not felt sufficient to justify government action. Given the immensity of Lake Superior, it was difficult to produce evidence of actual ecological damage by a single discharger. Laboratory tests indicated a possible impact on algae, but "these effects could not be easily measured in the lake even if they existed because any changes would be slight, subtle, and affected by other variables." Hence, except for the existence of the "green water," the "scientific case against Reserve was largely inferential and circumstantial."[15]

Reserve vehemently denied that it was causing any harm to the lake and adamantly refused to contemplate any change in behavior. After a series of meetings failed to produce any resolution, the chair of the federal enforcement conference (a now-obsolete mechanism) recommended that the Environmental Protection Agency (EPA) file suit because of Reserve's uncooperative attitude. The EPA was joined as a plaintiff by Minnesota and the other states bordering the lake. In the end, Reserve was to pay dearly for its intransigence.

The Trial

The case was filed before Judge Miles Lord. When I moved to Minnesota, Judge Lord was still on the bench, and I soon

14. See Robert Bartlett, *The Reserve Mining Controversy: Science, Technology, and Environmental Quality* 64–65, 70–72 (1980).
15. Id. at 77.

learned his reputation with local lawyers. Perhaps the best way of capturing Judge Lord's judicial reputation is to say that by comparison, the arch-liberal Justice William O. Douglas was considered a hair-splitting legalist. In the view of many lawyers, Judge Lord was on the bench to do justice and didn't allow anything to stand in the way—whether it was Congress, appellate courts, or the evidence in a case. Judge Lord's temperament is nicely captured by a story he once told a group of lawyers, which recalled how he got an additional doorway added to his conference room when he was a government attorney. After trying for months to get approval of the doorway from the General Services Administration (GSA, the agency that runs federal buildings), Judge Lord said he used a sledgehammer one weekend to break a hole in the wall. He then called GSA and told them, "When you fix that hole, remember to put a door there."[16]

This attitude—which produced the nickname "Miles the Lord"—led him into continual conflicts with the Eighth Circuit, in *Reserve Mining* as well as later cases.[17] Nevertheless, until he lost his temper late in the proceedings, Judge Lord's writings in *Reserve Mining* gave every impression of being thorough and well considered.

Judge Lord's conduct of *Reserve Mining* became something of a model for what has been called "public law litigation." He opened the suit to numerous interveners in order to obtain greater input.[18] He also brought in a number of experts who reported to him, rather than to the parties. As we will see, these experts had an important impact on the ultimate resolution of the case.

Until just before the trial, the dispute had centered on ecological damage to Lake Superior. But the focus of the case switched radically, almost by accident. The events are recounted in a book about the case:

16. Thomas Bastow, *This Vast Pollution* 141–142 (1986).

17. See Carol Rieger, "The Judicial Council's Reform and Judicial Conduct and Disability Act: Will Judges Judge Judges?" 37 *Emory L.J.* 45, 63–69, 71, 74, 77 (1988) (discussing Judge Lord's conduct in the Dalkon Shield cases).

18. See Richard Marcus, "Public Law Litigation and Legal Scholarship," 21 *U. Mich. J.L. Ref.* 647, 663–664 (1988).

[I]n May 1973, Glass, the EPA scientist with the Duluth National Water Quality Laboratory [NWQL], had a dream—a bad dream. He awoke with a fear of drinking water from Lake Superior. The next day, Glass explained his vision to Dr. Phillip Cook, a colleague at the NWQL. Cook, prompted by this vision and the alarm sounded by Mrs. Lehto [a Duluth resident who had heard about asbestos at a hearing], initiated a search for asbestos in Duluth's water supply. He found it.[19]

The EPA immediately went to Judge Lord with this new evidence; the asbestos data was made public, causing a panic in Duluth,[20] and the focus of the case switched from ecology to public health. The critical question was whether the asbestos posed a threat to the two hundred thousand people whose drinking water came from the lake.[21] This shift in emphasis clearly worked in favor of the government. As the court of appeals would later make clear, it considered public health a much more urgent concern than ecology.[22]

The trial took 139 days. The evidence included testimony from over a hundred witnesses, over fifteen hundred exhibits, and eighteen thousand pages of transcript.

Reserve initially claimed that the tailings did not contain asbestos, that they settled to the bottom of the lake without causing any pollution, and that any asbestos fibers found in the lake must come from natural sources. Judge Lord heard extensive testimony about such matters as the chemistry of asbestos, the behavior of the Lake Superior thermocline, and erosion patterns on the North Shore. For example, asbestos is the name, not of a particular compound, but of a family of minerals; experts debated whether the minerals contained in the taconite belonged in the same grouping as the minerals causing cancer among insulation workers.[23]

19. Schaumberg, *supra* note 9, at 149.
20. Id. at 150.
21. 514 F.2d at 517 & n.50.
22. Id. at 538.
23. 380 F. Supp. at 15, 31–33. For general background on asbestos and its health hazards, see James Alleman and Brooke Mossman, "Asbestos Revisited," *Scientific American,* July 1997, at 70.

We often talk loosely about environmental regulation being on the "frontiers of science," but this was dramatically so in *Reserve Mining*. The technology of the time was barely up to the task of identifying, let alone accurately counting, small asbestos fibers. For that reason, the judge faced great difficulty in determining the asbestos level in Duluth's drinking water.[24] Today we are often unsure whether a substance is a carcinogen or how the risk varies with the dose. For Judge Lord, however, even knowing the dose was a problem: fiber counts by different labs varied by a factor of ten. He concluded that the fiber count in Duluth drinking water ranged from 12 million fibers per liter up to perhaps 100 million at times.[25] Even the lower figure exceeds the EPA's maximum level for drinking water today.[26]

By the time of the *Reserve Mining* trial, it was known that inhaled asbestos is a very serious carcinogen. One issue in the case involved the effect of airborne asbestos on Reserve's workers and their families. There seemed to be little doubt that this health hazard required stringent pollution control. But even with respect to the air pollution, there were significant unknowns. The level of airborne asbestos was much lower than the level for insulation workers, and it was unclear how much of Reserve's airborne asbestos belonged to the most dangerous variety. Nevertheless, given the strong evidence of the harmfulness of airborne asbestos, the court of appeals found this part of the case relatively clear-cut.[27]

For present purposes, however, the most relevant portion of the case relates to water pollution. Here, even after asbestos counts in Duluth water were established and traced to Reserve's discharge, a major problem remained: nobody knew whether *drinking* asbestos was dangerous, and there was actually evidence it was perfectly safe. The evidence on this central issue will be discussed in connection with the Eighth Circuit's opinion on appeal.

24. 380 F. Supp. at 47–48.
25. Id. at 48.
26. See 40 C.F.R. § 141.51 (1997).
27. See 514 F.2d at 509–513. Because the court found the air emissions to present a more immediate hazard, the company was ordered to undertake in-

Judge Lord placed the burden of proof on Reserve to prove the safety of the current level of asbestos in Duluth's drinking water. Given the state of scientific uncertainty, the burden of proof was crucial to the outcome. The crux of his opinion is contained in the following passage:

> Defendants are exposing thousands to significant quantities of a known human carcinogen. If there is such a thing as a safe level of exposure to this human carcinogen, it must be very low and there is no credible evidence before this Court to indicate what that level is. Nonetheless the Court is asked to permit the present discharge until such a time as it can be established that it has actually resulted in death to a statistically significant number of people. The sanctity of human life is of too great value to this Court to permit such a thing.[28]

This left Judge Lord only with the question of remedy. Until late in the proceedings, Judge Lord apparently had hoped to find a solution that would not disrupt Reserve's operations. By the end of the trial, however, he had lost patience with Reserve. Reserve had filed misleading discovery responses and had also presented inaccurate, if not dishonest, testimony about the feasibility of dumping the tailings on land. Given the company's intransigence, Judge Lord saw no alternative but an immediate shut-down order, which he issued on April 20, 1974. If Reserve had taken a more cooperative attitude, Judge Lord probably would have given it time to put land disposal into effect. But he was apparently too outraged by Reserve's litigation tactics to allow any delay.[29]

The Appeal

Judge Lord's shut-down order was almost immediately stayed by the Eighth Circuit. The judges were all out of town for a judicial

---

terim control measures until the new land disposal system could be implemented. Id. at 539.

28. 380 F. Supp. at 54.

29. Id. at 20, 64–68, 83–84.

conference, so the arguments were held in a motel room, with the judges apparently in rather casual dress.[30]

The key to the stay order was the issue of risk. The court of appeals concluded that the evidence "does not support a finding of substantial danger" and that the discharges represented at most a "possible medical danger." Given the fact that the actual level of risk was simply unknown, the appellate court said, Judge Lord's "determination to resolve all doubts in favor of health safety represents a legislative policy judgment, not a judicial one." Although it expected Reserve to prevail on the health hazard issue, the appellate court did believe that Reserve's pollution of Lake Superior would eventually have to be abated.[31]

The government immediately asked the Supreme Court to vacate the Eighth Circuit order.[32] Only Justice Douglas voted in favor of the request to vacate the stay.[33] Proceedings immediately began before Judge Lord to identify a feasible method of land disposal.[34] Reserve initially planned to dispose of the tailings near the lake, but this plan was rejected, primarily because of the possibility that the dam would fail, spilling tons of tailings into the lake.[35] In the meantime, the appeal was briefed, and oral argument took place on December 9, 1974. The court's opinion was issued on March 14, 1975.[36] By then, the trial court was immersed in disputes about a new Reserve plan to dispose of the tailings at a site known as Milepost 7.[37]

The Eighth Circuit's decision, written by Judge Mike Bright, focused on the unclear possibility of a serious health hazard. As mentioned earlier, everyone knew that inhaling asbestos is dangerous. The question was whether drinking asbestos is also hazardous—and, indeed, whether ingested asbestos even enters the body at all. Ingesting asbestos presumably doesn't provide any

30. Schaumberg, *supra* note 9, at 194.
31. *Reserve Mining Co. v. United States,* 498 F.2d 1073, 1077–1078, 1083–1084 (8th Cir. 1974).
32. See 514 F.2d at 503–504 (reviewing history of stay requests).
33. *Minnesota v. Reserve Mining Co.,* 418 U.S. 911 (1974).
34. 514 F.2d at 504.
35. 380 F. Supp. at 79–80.
36. 514 F.2d at 492.
37. Id. at 506. See also Schaumberg, *supra* note 9, at 200–201, 221–224.

health benefits—so far as I know, no one has ever recommended asbestos as a source of dietary fiber—but there was genuine doubt about whether it was harmful. Much of the evidence failed to support fears of a cancer threat from the drinking water, and the contrary evidence, while significant, was more suggestive than compelling.

The Eighth Circuit carefully summarized the evidence bearing on the risk of waterborne asbestos. This evidence fell into three major groups: animal studies, studies of Duluth residents, and inferences from studies of asbestos workers.[38]

The animal studies were designed to determine whether asbestos fibers can penetrate the intestinal tract and enter the body. The results of the studies were conflicting as to whether asbestos can do so. In none of the studies was there evidence that asbestos caused cancer in lab animals.[39]

Today animal tests are a familiar part of investigation of possible carcinogens. Usually, the problem in interpreting the tests is rather different from that posed in *Reserve Mining*. Frequently, when positive results are obtained in animal tests, it is difficult to extrapolate the findings to humans. Different species have different responses to chemicals. The same substance may be a carcinogen in mice but not in rats, or in rats but not in humans (or, for that matter, in humans but not in rodents.) Moreover, to obtain observable results with small groups of rodents, very high doses are needed. But then it is difficult to project from these large doses to the effects of much smaller doses over extended periods.[40] Because the animal tests in *Reserve Mining* were resoundingly negative—often failing to show any absorption, let alone tumors—these interpretative questions did not have to be faced.

Although these animal tests tended to clear Reserve, they could not be considered conclusive. For instance, benzene is one

38. 514 F.2d at 514–519.

39. Id. at 515–516.

40. For an excellent discussion of the difficulties of interpreting animal studies, see John Graham et al., *In Search of Safety: Chemicals and Cancer Risk* 39–63, 140–148, 167–172, 181–183 (1988). The "Symposium on Risk Assessment in the Federal Government," 3 *N.Y.U. Envtl. L.J.* 251–258 (1995), provides extensive background information on this and other issues.

of the best-documented human carcinogens, but until recently, it has been impossible to detect any effect in animal tests.[41]

The next category of evidence involved studies of Duluth residents. One crucial study was initiated by Judge Lord himself. The purpose of the study was to determine whether the tissues of long-time residents contained asbestos fibers. Tissue samples from Duluth cadavers were compared with samples from Houston, which does not have asbestos in its drinking water. The plaintiffs predicted that the study would yield solid information about the risk to Duluth residents. When the study was completed, however, the asbestos fibers failed to appear. The court of appeals found this result highly significant and was not impressed by efforts to explain away the negative findings. As the court also noted, epidemiological studies had failed to produce evidence of an elevated gastrointestinal cancer rate in Duluth.[42]

The failure of the epidemiological studies to produce evidence of harm was not necessarily surprising. It might seem easy to study two groups with different levels of exposure, compare disease rates, and determine if a substance is dangerous. In reality, the studies are difficult and expensive. People move in and out of urban areas, making it hard to track the exposed population; doses are difficult to reconstruct after the fact; medical diagnoses are not necessarily written up with the needs of future researchers in mind; and a host of confounding variables such as different rates of smoking need to be controlled for. Given all these difficulties, relatively small effects are quite difficult to detect and tend to be lost in the background "noise" caused by other variables and measurement errors. As one careful study of risk assessment methods explains:

> Even if historical exposure data are accurate and confounders nonexistent, epidemiological methods generally cannot detect 1 in 1000 to 1 in 1,000,000 lifetime risks. . . .

41. Graham et al., *supra* note 40, at 116–117.
42. 514 F.2d at 514–515, 518 n.50. A 1980 study, however, did find increased numbers of asbestos fibers in Duluth tissue samples as compared with Houston and St. Paul, Minnesota. Criteria and Standards Div., U.S. Envtl. Protection Agency, *Ambient Water Quality for Asbestos* C-28 (Oct. 1980).

The limited capacity of epidemiology to detect effects means that negative results are not unlikely, even when the chemical is a genuine human carcinogen. This possibility further confuses the interpretation of any studies that are available. Are negative results "real" in this case, or are they the result of low statistical power?[43]

The authors agree that "epidemiological findings can sometimes place a useful upper bound on excess human cancer risk," but they warn that "these upper bounds will generally be above the minimum risk levels of concern to regulators." Thus, it would probably be a mistake to put too much weight on the negative epidemiological studies. At best, assuming no methodological problems, they only proved that the asbestos would probably not kill any more than about a hundred of Duluth's 100,000 residents.

Up to this point, the court had found virtually no evidence of harm. The really worrisome point, however, was that asbestos workers do have a significantly increased rate of gastrointestinal cancer. A plausible explanation is that they cough up asbestos fibers and then swallow them, resulting in gastrointestinal cancer. Dr. Selikoff, the leading expert on asbestos-related disease, considered this a likely mechanism. Assuming that to be correct, another expert compared the total exposure of Duluth residents to gastrointestinal exposure levels of asbestos workers. Based on this testimony, Judge Lord found the risks to be of similar magnitude.[44]

The comparable levels of occupational and environmental exposure eliminated the need for the court to consider one of the most vexing problems in risk assessment. Usually, harmful effects are observed in occupational exposures that are far above the normal environmental level. Consequently, it is necessary to extrapolate the effects of high-level exposure to much lower level exposures. This extrapolation requires a "dose-response" curve showing how the harmful effects relate to dose. There are various statistical models for doing this, all of which more or less fit

43. Graham et al., *supra* note 40, at 181.
44. See 514 F.2d at 516–517 (reviewing Judge Lord's findings and the supporting evidence).

the data, are open to debate, and yield very different results. To make matters even more confusing, it is not even clear that the same curve applies to all carcinogens. Different mechanisms of cancer causation may correspond to different curves. The EPA's response to these problems is to base risk assessments on what is called the "upper confidence level of the multistage model."[45] For our purposes, the meaning of these technical terms is of little concern. The important points are that the model produces risk estimates that vary in direct proportion to the size of low doses and that most scientists view this as a "conservative" method, which is likely often to overestimate actual risks.[46] Fortunately, Judge Lord was able to bypass this entire issue, given the exposure estimates.

In its order staying Judge Lord's injunction, the court of appeals seemed skeptical of the health claims. As the later opinion on the merits made clear, however, the appellate judges were actually quite concerned about the health risks. Although they found the evidence too inconclusive to justify an immediate shut-down, they also seemed reluctant to take the chance of contributing to what might later turn out to be a public health disaster.

In evaluating the evidence, the Eighth Circuit was obviously impressed by the testimony of another expert, Dr. Arthur Brown, who was quoted at length twice in the text of the opinion.[47] Dr. Brown chaired the pathology department at the Mayo Clinic and had served as a technical advisor to Judge Lord and as an impartial witness. Speaking as a scientist, Dr. Brown said that the evidence "is not complete in terms of allowing me to draw a conclusion one way or another concerning the problem

45. The details are explained in Graham et al., *supra* note 40, at 41–47, 155–164.

46. For a defense of the EPA's method, see Adam Finkel, "A Second Opinion on an Environmental Misdiagnosis: The Risky Prescriptions of Breaking the Vicious Circle," 3 *N.Y.U. Envtl. L.J.* 295, 340–352 (1995). Despite the notorious willingness of law professors to stake out confident positions regarding technical disputes in other fields, I have no intention of expressing any view about this scientific controversy.

47. 514 F.2d at 513–514 (Dr. Brown's testimony concerning air pollution); id. at 518–519 (his testimony on the water pollution issue).

of a public health hazard in the water in Lake Superior." Speaking as a physician, however, he had more definite views:

> As a medical person, sir, I think that I have to err, if I do, on the side of what is best for the greatest number. And having concluded or having come to the conclusions that I have given you, the carcinogenicity of asbestos, I can come to no conclusion, sir, other than that the fibers should not be present in the drinking water of the people of the North Shore.[48]

The appellate court's own conclusions were quite similar. According to the court, "[I]t cannot be said that the probability of harm is more likely than not. . . . On this record it cannot be forecast that the rates of cancer will increase." The most that could be said was that asbestos contamination in Lake Superior "gives rise to a reasonable medical concern for the public health." This concern was enough, however, to justify what the court called "abatement of the health hazard on reasonable terms."[49] In reaching this conclusion, the court also seemed influenced by its finding that Reserve was in violation of its state permit, so that the discharge was not only a possible threat to public health, but also unlawful.[50]

The remaining issue was the remedy. The court believed that an immediate shut-down was untenable, given the inconclusive evidence of any danger to the public. A shut-down would not simply have harmed shareholders. There were other stakeholders as well. Reserve had over three thousand employees, each of whom supported between four and six other people (directly or indirectly), and the firm produced over 10 percent of the nation's iron. Moreover, the union argued that the health effects of a plant closure on workers might be more severe than those caused by the asbestos. (As it turns out, the union's argument was not without foundation. There is reason to think that unemployment, and perhaps also less drastic reductions in income, do pose a substantial health hazard comparable in seriousness to

48. Id. at 506 n.18, 518–519.
49. Id. at 520.
50. Id. at 531.

those from environmental carcinogens.[51] One estimate is that for every $37 million spent to comply with a regulation, an additional statistical death results.[52]) At oral argument, Reserve said it was willing to spend $243 million to halt its air and water pollution. Based on all these factors, the court concluded that Judge Lord had abused his discretion by ordering an immediate shutdown and that the company should be given a reasonable time to switch to land disposal.[53]

The Justice Department lawyers who handled the case were appalled by the decision. Given the court's assessment of the risk, they felt that the remedy was weak and essentially empty. When the head of the EPA endorsed the court's decision, the lawyers considered his action a "craven crawl," surrendering principle to expediency.[54]

Aftermath

Judge Lord had made no secret of his displeasure with the stay order.[55] In the closing portion of its opinion, the appellate court rebuked him for taking actions "which appear to be in conflict with the express language" of the stay.[56] After the court of appeals' opinion on the merits was announced, Judge Lord became even more outspoken. At one point, he hauled a number of company officers and state officials into court to hear a lecture—

51. See W. Kip Viscusi, "Risk-Risk Analysis," 8 *J. Risk & Uncertainty* 5 (1994). Viscusi has estimated that there is a loss of one statistical life for every $50 million drop in GDP. W. Kip Viscusi, "Equivalent Frames of Reference for Judging Risk Regulation Policy," 3 *N.Y.U. Envtl. L.J.* 431, 458 (1994). For a critique of this theory, see Thomas McGarity, "A Cost-Benefit State," 50 *Admin. L. Rev.* 7, 46–49 (1998).

52. W. Kip Viscusi, "Regulating the Regulators," 63 *U. Chi. L. Rev.* 1423, 1455 (1996).

53. 514 F.2d at 536–538.

54. Bastow, *supra* note 16, at 182–183.

55. After expressing "confusion" regarding the Eighth Circuit's instructions on remand, *United States v. Reserve Mining Co.,* 380 F. Supp. 11, 71–72 (D. Minn. 1974), Judge Lord proceeded to dissect Reserve's proposed plan for land disposal. Id. at 72–86. He concluded with an affirmation of his decision to enjoin Reserve's continued dumping and a not-so-veiled attack on the court of appeals' stay order. Id. at 89–91.

56. 514 F.2d at 541.

perhaps sermon would be a better word—about the danger of asbestos pollution.[57] The court of appeals then removed him from the case for bias, violation of the company's due process rights, and disregard of the Eighth Circuit's mandate. Judge Lord's unconcealed disdain for the appellate court received the following response:

> Disregard of this court's mandate by a lawyer would be contemptuous; it can hardly be excused when the reckless action emanates from a judicial officer. It is one thing for a district judge to disagree on a legal basis with a judgment of this court. It is quite another to openly challenge the court's ruling and attempt to discredit the integrity of the judgment in the eyes of the public.[58]

Judge Lord was replaced by Judge Devitt, a considerably less colorful, if "sounder," jurist. (After further clashes with the court of appeals in other cases, Judge Lord would later leave the bench to become a personal injury lawyer. His advertisements advise accident victims to dial 333-LORD.)

The court of appeals may have thought it had set the stage for a prompt resolution of the case. In reality, protracted litigation in the state courts was yet to come over the location of the land disposal site. In the spring of 1977, the Minnesota Supreme Court issued an opinion approving the Milepost 7 plan.[59] Further litigation followed about the conditions of the permit, which resulted in another Minnesota Supreme Court decision a year later.[60] Reserve then announced that it was proceeding with its conversion to land disposal.[61] The conversion was completed in 1980.[62]

Ironically, the issue of waste disposal soon became rather academic for Reserve. In June of 1982, the company temporarily shut down, blaming deterioration in the American steel industry.

57. *Reserve Mining Co. v. Lord,* 529 F.2d 181, 187 (8th Cir. 1976).
58. Id. at 188–189.
59. *Reserve Mining Co. v. Herbst,* 256 N.W.2d 808 (Minn. 1977).
60. *Reserve Mining Co. v. Pollution Control Agency,* 267 N.W.2d 270 (Minn. 1978).
61. "Reserve to Continue Mining," *Facts on File* 553, at E2 (1978).
62. Patrick Marx, "Minn. Mining Firm Ends Dumping in Lake Superior," *Wash. Post,* Mar. 18, 1980, at A24.

It reopened six months later, but then closed for another six months. The company never did regain its financial footing. In 1986, LTV (by then one of the co-owners) filed bankruptcy under Chapter 11 and withdrew from the partnership. Shortly thereafter, Armco (the other partner) announced it was closing Reserve's operations. By August 1988, the Armco subsidiary running Reserve had also filed under Chapter 11. In 1989, Reserve was sold for $52 million to a firm called Cyprus Minerals Co., which resumed operations on a greatly reduced scale.[63]

This later financial history naturally raises the question of whether land disposal bankrupted the company. Apparently, however, the blame was placed on declines in the U.S. steel industry, rather than on any special problems at Reserve. It seems unfair to criticize the 1975 judicial ruling for apparently unforeseeable economic developments. The courts in 1975 were faced with an economically healthy defendant; they had little reason to know that America's heavy industry (including the steel industry) faced a crisis.

To the extent that the company's financial troubles might have been foreseeable, their import is unclear. Extractive industries like mining are plagued by unstable demand, regular periods of oversupply, and low value-added. Arguably, the court should have given less weight to possible economic dislocations because of these industry characteristics. A recent economic study of conflicts between the environment and the extractive industries concludes as follows:

> Most extractive activities produce relatively uniform commodities that are readily available from other sources. Oversupply is the reason agricultural, fuel, metal, and fiber prices are so low and the reason these industries are in relative decline. Modest increases in

---

63. "Reserve Closure," 298 *Mining J.* 457 (1982); "Reserve Mining to Close," 300 *Mining J.* 211 (1983); Thomas Hayes, "LTV Corp. Files for Bankruptcy; Debt Is $4 Billion," *N.Y. Times,* July 18, 1986, at A1; "This Month in Mining: Minnesota," *E&MJ: Engineering and Mining J.,* Oct. 1986, at 134; "Reserve Mining to Re-open, Says Cyprus Minerals," *E&MJ: Engineering and Mining J.,* July 1989, at 16N.

their production will add relatively little economic value because of limited demand for the product.[64]

The report warns that policies adopted to save declining industries are unlikely to succeed and adds that "if we opt for extractive activity to keep the local economy afloat, we will be sacrificing what is scarce and unique for what is common and cheap."[65] Be this as it may, the taconite industry has recently recovered, and Reserve's old operation is currently thriving under new ownership, at least for the moment.[66]

THE COURTS IN *Reserve Mining* were faced with a difficult task in weighing an uncertain risk to public health against an approximately $200 million expenditure. The question posed by the case is how to go about making such trade-offs between safety and cost. *Reserve Mining* provides an excellent context in which to discuss why these choices are so hard, what methods have been proposed for making them, and how we might go about resolving these cases in a sensible way.

Reserve Mining is a difficult case for several reasons, even apart from the scientific uncertainly about the extent of the risk. The first is the difficulty of somehow assessing the weight to give economic costs versus a possible public health risk. The values at stake are so different that we aren't sure how to compare them. Chapters 2 through 4 explore that issue, arguing that both values deserve weight, but that we should begin with an environmentalist baseline. The analytic difficulties are compounded by the temporal dimension of environmental problems. The risks facing the appeals court were very long-term, and it is hard to weigh risks that may not materialize for decades against money and jobs that will be lost today. Also, during the dispute, there were rapid changes in scientific knowledge—the asbestos threat to drinking water wasn't even known when the case began. We

64. Thomas Power, *Lost Landscapes and Failed Economies: The Search for a Value of Place* 237 (1996).

65. Id. at 235, 254.

66. See Jane Brissett, "Iron Willed: Workers and Management Save a Taconite Plant, and a Way of Life," *Corp. Rep. Minn.*, Sept. 1, 1996, at 44, 48.

can expect future scientific knowledge to continue to change in dramatic and unexpected ways. Our level of ignorance, in other words, is also subject to future change. Chapters 5 and 6 consider the implications of these timing issues for environmental decisions.

CHAPTER TWO

Economics versus Politics

The most profound issue raised by *Reserve Mining* is how we should think about the conflicting values at stake in the case. We might say that environmental values such as risks to human life and to the lake's ecology are in a wholly different class than the mere expense of correcting the problem. This was Judge Miles Lord's view, at least by the end of the case. It led him to order an immediate shut-down of the plant in order to serve the paramount good of public health. Alternatively, we might view the environmental values as significant only to the extent they can be translated into financial terms. If so, the way to make a decision is a cost-benefit analysis.

Much of the environmental scholarship of the past twenty years has been dominated by the struggle between these opposing viewpoints.[1] Many environmentalists consider the economic approach crass and blind to deeper values. They are fond of the saying that an economist is someone who knows the price of everything, but the value of nothing. Economists consider environmentalists to be hopeless romantics, eager to pursue their own personal values without heeding the cost to society. They find it easy to associate environmentalism with the image of Don

---

1. For a recent overview of the debate, see David Driesen, "The Societal Cost of Environmental Regulation: Beyond Administrative Cost—Beyond Analysis," 24 *Ecology L. Q.* 545 (1997).

Quixote tilting at windmills. Thoughtful members of both groups, of course, temper these exaggerated responses, but the conflict remains quite real. Any perspective on environmental law must either take sides or find a way of mediating the dispute.

This conflict is often approached as if it involved two completely different visions of life: one in which the only goal is the fulfillment of human desires; the other in which the exclusive goal is the well-being of the planet as a whole. But at least within the mainstream of environmental scholarship, it would be hard to find anyone who takes these extreme positions. Few, if any, serious scholars would dismiss the intrinsic value of preserving the Grand Canyon or the redwoods; even fewer would dismiss saving human lives as wholly unimportant compared with saving wilderness. Nevertheless, the conflict is a serious one. Once we dismiss the more extremist positions, we are still left with the question whether cost-benefit analysis is the key to sound environmental policy or a largely irrelevant (if not deceptive) distraction.

Critiques of cost-benefit analysis often turn on the deficiencies of its underlying measure of value, the willingness of consumers to pay for goods and services. These critics argue that how much people are willing to pay for something proves nothing about its moral value. As we will see, they have good reason to reject the idea that "willingness to pay" is definitive of value. Their critiques go wrong, however, in arguing that willingness to pay is a completely irrelevant factor in assessing the public interest.

Putting aside the details of *Reserve Mining* for the moment, this chapter analyzes these conundrums. I will begin by attempting to describe the two conflicting positions in more detail. Much of the dispute involves the concept of economic efficiency, which is based on "willingness to pay" as a standard of value. Before analyzing the attacks on this standard, it behooves us to consider carefully how economists define economic efficiency and how they go about valuing intangibles like human life and environmental quality. This sets the stage for a discussion of the attacks on economic efficiency as a legitimate social goal and, finally, for an analysis of the true relevance of economic efficiency to environmental issues.

Our society has two principal methods of making decisions: politics and the market. The standard of economic efficiency is drawn from the marketplace. Some economists view this market-related standard as supreme. Their critics view it as irrelevant to public policy, which they would base solely on the political process. In reality, both politics and markets express the values of the public, and both sources of information about societal values deserve consideration in formulating public policy. Thus, economic efficiency should be regarded as neither controlling (as some economists would say) nor irrelevant (as some of their critics maintain).

An Overview of the Dispute

One approach to *Reserve Mining* is embodied by Judge Lord's refusal to countenance a continuing health risk, regardless of cost. In contrast, a cost-benefit analysis compares health benefits with economic costs. For example, we could calculate the value of the lives saved by eliminating asbestos from the drinking water. To determine the economic value of life, we could use labor markets to find how much money workers demand in exchange for riskier occupations. Perhaps we could also try to calculate the value of the integrity of the lake itself. We might, for example, conduct a national survey asking people how much they would pay to prevent dumping in the lake. Then we could compare these numbers with the expense of land disposal. We will consider some of these assessment methods in more detail later; for present purposes, it is only the overall approach that is important.

The economic approach has strong support among scholars, political leaders like Newt Gingrich, and some judges. Although neglected in *Reserve Mining,* it was embraced with enthusiasm in a later asbestos case, *Corrosion Proof Fittings v. EPA.*[2] After intensive investigation, the EPA had decided to use its powers under the Toxic Substances Control Act to ban most asbestos products. It estimated that the rule would save either 202 or 148 lives (depending on whether future deaths are counted equally with

2. 947 F.2d 1201 (5th Cir. 1991).

present ones) at a cost of $450–800 million (depending on the
price of substitutes for asbestos). Perhaps because of annoyance
that the agency had failed to consider less drastic remedies, the
Fifth Circuit brusquely overturned the ban, finding a variety of
flaws in the EPA's analysis.

In *Corrosion Proof Fittings,* the court came out strongly in favor
of quantitative economic analysis. For example, the court
faulted the EPA for rejecting less stringent regulations without
calculating costs and benefits: "[w]ithout doing this it is impossible, both for the EPA and for this court on review, to know that
none of these alternatives was less burdensome than the ban in
fact chosen by the agency." Later in the opinion, the court con-
ceded that unquantified benefits need not be completely ex-
cluded from the EPA's analysis. Still, the court gave such benefits
only limited weight. The concept of unquantified benefits, the
court said, "is intended to allow EPA to provide a rightful place
for any remaining benefits that are impossible to quantify after
the EPA's best attempt, but which still are of some concern." But
unquantified factors can play only a minor role; they "can, at
times, permissibly tip the balance in close cases," but they cannot
"be used to effect a wholesale shift on the balance beam."[3]

Corrosion Proof Fittings has gotten a very mixed reception. It
receives favorable mention from advocates of "regulatory reform" like Justice Stephen Breyer[4] and Cass Sunstein, a leading
scholar of administrative and constitutional law at the University
of Chicago.[5] What writers such as Sunstein consider a "promising model for the future,"[6] environmentalists consider an abomi-
nation. Tom McGarity, an environmental law specialist at the
University of Texas, accuses the court of misinterpreting the

3. Id. at 1208, 1217, 1219. The court also criticized the agency's handling of
numerous specific issues: the method used to assign present values to future
deaths, failure to demonstrate the existence of safe alternatives to asbestos in
various uses, and prohibition of some products posing very small risks (which
was the source of Justice Breyer's example about "ingested toothpicks"). Id. at
1218–1220, 1223 & n.23, 1226–1227.

4. Stephen Breyer, *Breaking the Vicious Circle: Toward Effective Risk Regulation*
14 (1993).

5. Cass Sunstein, "Legislative Foreword: Congress, Constitutional Moments,
and the Cost-Benefit State," 48 *Stan. L. Rev.* 247, 294 n.235 (1996).

6. Cass Sunstein, "Health-Health Tradeoffs," 63 *U. Chi. L. Rev.* 1533, 1566
(1996). Sunstein does, however, criticize the court for being too hostile toward

statute, of offering gratuitous wisdom on technical issues beyond its competence, and in general of sending the "EPA on a potentially endless analytical crusade in search of the holy grail of *the* least burdensome alternative."[7] Since the case was decided, the EPA has apparently given up entirely on attempting to implement the underlying statute, which has become a dead letter.

Outside the courts, the contrast between conflicting views is even starker. At the risk of caricature, the choice seems to lie between "tree huggers," who hold the environment sacred and reject economic values as profane, and "bean counters," who believe only in values that can be quantified in dollars and cents.

Both perspectives strike me as unbalanced. The tree-hugger side of the argument unduly elevates so-called public values over mere private preferences, putting the environment in one category and economic interests in the other. This analysis rests on too facile a dichotomy between the economic and the political realms. But the bean-counter side of the argument is no more satisfying. It misconceives the appropriate role for cost-benefit analysis by acting as if economic efficiency were a sort of "super value" that measures all other values. Though we should take economic interests into account when making environmental decisions, the bean counters are wrong to view cost-benefit analysis as *the* method of making environmental decisions.

As with many disputes, the hardest problem may be not deciding which side is right, but finding a framework in which to evaluate the opposing arguments. One way to conceptualize cost-benefit analysis is to consider economic efficiency a standard of social welfare. This argument runs through the following implicit steps:

(1) The fundamental moral value is human welfare.
(2) Individual welfare means the satisfaction of an individual's preferences to the greatest degree possible.

---

regulations designed to force the creation of new technologies. Id. at 1554 & n.71, 1566 n.121.

7. Thomas McGarity, "The Courts and the Ossification of Rulemaking: A Response to Professor Seidenfeld," 75 *Tex. L. Rev.* 525, 541–549 (1997). See also Lisa Heinzerling, "Political Science," 62 *U. Chi. L. Rev.* 449, 463–464 (1995) (sharply criticizing Justice Breyer's discussion of the case).

(3) Group welfare means the highest collective level of preference satisfaction, using cost-benefit analysis to determine when we have succeeded in increasing that level.

Under this view, economic efficiency is good in and of itself as a
measure of social welfare.

Conceptualized in this way, the dispute proceeds in highly abstract terms. We might ask, for instance, whether moral values
go beyond the needs and interests of humans, so that the environment has an intrinsic value apart from its effect on human
welfare. Or we might think of the dispute as relating to the
proper way of thinking about moral decisions. Should we think
of them, as economists do, as problems in maximizing satisfaction of preferences? Or are moral decisions exercises in some
noninstrumental form of reasoning? Questions such as these go
to the heart of moral philosophy.

I am doubtful that this is a fruitful way to approach the debate
over cost-benefit analysis. It lands us immediately in very deep
philosophical waters, forcing us to debate the fundamental nature of ethics before we can get to the specific issue of cost-
benefit analysis. In effect, we are being given the following recipe
for deciding environmental policy issues: "*Step 1:* Settle the
question originally raised by Plato by providing an indisputable
definition of the nature of 'the good.' *Step 2:* Apply the results of
step 1 to the particular problem of environmental quality." It's
easy to see that we're unlikely to get past step 1 anytime soon.

Moreover, this way of conceptualizing cost-benefit analysis is
something of a straw man. No one really believes that what is
good for a person is *defined* by what he or she happens to prefer
at a given moment or that what is good for society is *defined* by
an increase in the gross domestic product (GDP). What most
economists actually believe is that individual preferences are
generally a useful guide to what is good for people. In any event,
many economists also believe, people should normally be allowed to pursue their own views of what is good for them,
whether or not they are right in some cosmic sense. They also
tend to believe that the GDP is an indicator of whether people
are collectively better off, but it is impossible to present a coherent argument that GDP defines social welfare. For one thing,
the prices used to measure GDP (and to perform cost-benefit

analysis) are affected by government policies, so they provide an unstable metric for social welfare—a yardstick that potentially changes whenever we measure something new. Thus, approaching the dispute as a debate over the definition of moral values misses the economists' real point.

There is a more fruitful way to conceptualize the dispute. Rather than thinking about economic efficiency as something that is good in itself, we can think of cost-benefit analysis as a mechanism for resolving certain disputes about how to allocate resources. If several people want to own the same antique rocking chair, we often settle the matter by holding an auction so the one who is willing to pay the most gets the rocker. Assuming that wealth is distributed fairly (no small assumption!), the auction seems like a fair way to settle the question of who gets the chair. Everyone has a fair chance to participate, and someone who wants the chair more strongly can express that desire by bidding higher. If for some reason an auction is impractical, we might want to use some other technique such as cost-benefit analysis to decide who *would* have paid more if the auction had been held. In other words, we can view cost-benefit analysis as being a method for making social decisions, which sometimes has desirable qualities, like giving the rocker to the person who wants it most in comparison with other goods he or she might buy. The auction is essentially a precise way of learning how much money each person is willing to trade for the chair, a function performed by markets more generally.

If we think of cost-benefit analysis this way, then the dispute isn't over the intrinsic moral value of human life or endangered species, but rather about how we should determine the trade-offs that the public is willing to make. One method is to use markets, augmented by techniques such as cost-benefit analysis where markets are unable to function effectively. The other method is provided by the political process. The dispute between tree huggers and bean counters is over the legitimacy of the market-based method as a mechanism for social choice.

Consider a judge or an EPA official who must make a decision about the level of environmental protection in a situation where the law does not dictate the result. Her own environmental values may also be relevant to the decision, but let's assume she believes her main role is implementing the public's views. Her difficulty is that there are two different ways in which members of the public can express their views about the environment. On the one hand, they can engage in political activities: joining environmental or community groups, discussing issues, lobbying, and voting. Here, the ultimate test is political: "willingness to vote." On the other hand, they can also express views about the environment in their private lives, indicating their views by demanding more money for riskier jobs, traveling long distances to visit wilderness areas, or paying more for houses in communities with better environments. Here, "willingness to pay" is the test. The problem facing our hypothetical decision maker is that these two kinds of indicators of the public's views may or may not coincide. If not, she must decide how to deal with the conflict: Should she view politics or the market as the best indicator of the public interest? Or should she give some weight to each?

From this perspective, the question is what kinds of institutions better indicate the "public interest" in particular situations. Bean counters think market indicators should control the decision; tree huggers would disregard the market message entirely in favor of political indicators.

Willingness to pay or willingness to vote? Much of the discussion of environmental values has been dedicated to analyzing the flaws in one or the other position, as if demonstrating these flaws were enough to decide the issue. But in reality, both market and political mechanisms have flaws as expressions of the public interest. The real question is not whether to follow one while ignoring the other, but how to make the best use of both to guide public policy.

I will argue that willingness to pay cannot, in and of itself, determine the public interest. As a practical matter, the state of the art does not allow sufficiently precise economic measurements that could be given controlling weight in environmental decisions. In addition, there are compelling arguments against defining the public interest solely in terms of willingness to pay. But the arguments for completely disregarding economic signals about environmental quality are also shaky. There are sound reasons why willingness to pay is relevant to environmental decisions. Precisely what role it should play is another matter, which will be explored in chapters 3 and 4.

The dispute over cost-benefit analysis connects with a broader debate in legal scholarship. Although what I have called the tree-hugger perspective is often associated with environmentalism, it is also linked with a movement in legal scholarship called "neo-republicanism" (not to be confused with the GOP variety). Neo-republicans have argued that conventional legal and economic analyses are too individualistic. They argue that political institutions, unlike markets, function as forums for deliberating about collective values. Many of the neo-republicans are environmentalist stalwarts, but there are important exceptions such as Cass Sunstein.

The neo-republicans are partially right: when we are discussing such public goods as environmental quality, we necessarily must make the decisions through some collective political process because markets cannot determine the appropriate level of public goods.[8] Although the neo-republican critique of economics is one-sided, economic analysis of collective decisions does tend to overrate the significance of the kinds of preferences that are displayed in markets. Where the neo-republicans go wrong, I argue below, is in creating such a sharp dichotomy between what happens in the "public" setting of politics and what happens in the "private" setting of the market. Unless our society is going to be completely schizophrenic, the two must relate to each other. In particular, this means that economic preferences often deserve a hearing by policy makers.

Before we try to determine the proper role of economic analysis, we first need to understand how economists measure environmental values. I then turn to the neo-republican critique.

8. The key characteristics of public goods are joint supply and nonexcludability, which together mean that everybody essentially consumes the same amount of the good. Nonexcludability means that it is impractical to prevent individuals from consuming the goods once they are made available; joint supply means that it is more efficient to supply the goods jointly anyway because of low or nonexistent marginal cost. Classic examples are national defense and clean air, which can practicably be supplied only to the population as a whole. See Dennis Muehler, *Public Choice II* 10–11 (1989).

Finally, I discuss some reasons for thinking that economic preferences, although not determinative, are relevant to public policy.

The Economist's Perspective

Willingness to pay is a familiar standard in other contexts. For example, if you ask a real estate appraiser to determine the value of your house, he will try to determine what a seller would be willing to pay by looking at recent comparable sales. The resulting figure may or may not correspond to your personal feelings about the house, but it does tell you what the house is worth in the marketplace. Cost-benefit analysis applies the same standard to government regulations. For example, to determine the value of clean drinking water, we might ask how much people would be willing to pay for purer water. In this context, however, use of "willingness to pay" is less familiar, if not unsettling.

Why do economists adopt willingness to pay as the test for government regulations? The answer is rooted in the theory of welfare economics. To understand the arguments for the "willingness to pay" test, we need to delve into economic efficiency, the normative standard used in welfare economics. Unfortunately, doing so also requires that we confront some fairly technical concepts, not to mention obscure economic jargon. But we cannot fairly evaluate the economic approach without understanding its foundation.

Defining the Standard

The "gold standard" for welfare economists is the "Pareto improvement." One situation is a Pareto improvement over another if at least one person benefits from the change and no one is hurt. Essentially, no one would have any reason to vote against a Pareto improvement. For example, suppose everyone in Duluth would feel that they were better off with asbestos-free drinking water but $100 less in cash. Also suppose we were able to achieve this goal using purification equipment, which we could finance by charging each resident $50. This proposal would make everyone better off than they were to begin with, on their own reckoning, while harming no one—it would receive unanimous support over the status quo. This is a Pareto improvement.

Pareto improvements are hard to oppose—after all, no one is hurt, and someone is made better off, at least within his or her own frame of reference. If we think that generally people are good judges of their own interests, we can normally conclude that a Pareto improvement is good for them, at no cost to anyone else. This is as close to a free lunch as you can get in economics.

Unfortunately, like free lunches generally, Pareto improvements are not easy to find. It's hard to avoid causing at least a small amount of harm to someone in the course of benefitting others. We might try to compare the losses and gains through some kind of measure of personal happiness so we could decide if society as a whole is happier. But economists are skeptical about the meaningfulness or practicality of such interpersonal comparisons of emotional states. Instead, they have reformulated the concept of economic efficiency to deal with such situations, while avoiding the need for interpersonal comparisons of well-being.

Economic efficiency is a somewhat tricky concept. One situation is more economically efficient than another if the gains of the winners exceed the losses of the losers. More precisely, the economic efficiency standard is satisfied if, following a hypothetical transfer from the winners to the losers, the resulting situation would be a Pareto improvement over the original state of affairs. Or, more simply, the winners benefit enough that they would be willing, if necessary, to bribe the losers into going along with the change. In some general sense, economic efficiency tells us whether the total benefits of a change outweigh the total costs, though some people may win and others may lose. Thus, an economically efficient improvement would pass a cost-benefit analysis.

An example may clarify the meaning of the economic efficiency standard and how it relates to the Pareto standard. Consider again the situation of the Duluthians, who would be willing to pay $100 apiece (or a total of $10 million) for asbestos-free drinking water. If the water purification equipment costs $5 million, we saw earlier that a $50 per head tax to finance the equipment would be a Pareto improvement: the residents would give up $50 to obtain a benefit worth $100 to them. Now suppose we decide it is unfair to tax the residents, since Reserve Mining created the problem in the first place. Instead, the court orders Reserve to pay the $5 million to install the new equipment.

This can no longer be considered a Pareto improvement because some people (the owners of Reserve) are worse off, while others (the Duluthians) are better off. If the people of Duluth then reimbursed the company, we would have a Pareto improvement, since the residents would be better off than they were without the equipment and the company would be made whole (and therefore indifferent to the whole transaction). Without the actual reimbursement, we can still say that the benefit to the Duluthians is large enough that they *could* have afforded to reimburse Reserve. Hence, the court order is economically efficient. If we count the value of the improved drinking water as $100 per head, we have increased overall GDP. (We have also transferred wealth from the company to the residents, which raises an ethical rather than an economic issue of the kind covered by cost-benefit analysis.) Hence, a cost-benefit analysis would come out positive.[9]

9. One of the reasons for throwing this technical jargon into the discussion is to emphasize that we are dealing with a technical subject, not just a commonsense judgment about social welfare. Economic efficiency should not be confused with the commonsense notion of economic prosperity, although the two are loosely related. Economic efficiency does capture some aspects of the everyday concept of prosperity. If you increase the total economic value of all assets and activities—measured by how much people would be willing to pay for them, given the current distribution of wealth—then economic efficiency has been enhanced. So we could say, somewhat loosely, that economic efficiency "increases social wealth."

But the phrase "social wealth" is somewhat misleading. For instance, an improvement in economic efficiency might leave all physical and financial assets unchanged, but increase the total amount of leisure. Since people are willing to pay for leisure (as shown by their willingness to give up hours of paid work to enjoy it), leisure counts just as much as stocks and bonds, or houses and cars. Also, there really is no such entity as "social wealth." We're dealing with a comparison between two specific states of affairs, not with some objective quality that truly can be measured in absolute terms. Finally, economic efficiency ignores completely the question of how wealth is distributed, so "social wealth" could increase at the expense of plunging much of the population into desperate poverty. In short, cost-benefit analysis is based on a very technical, nonintuitive method of identifying economic improvements, which is sometimes a reasonable proxy for more commonsense notions.

Applying the Standard

In operational terms, economic efficiency requires that we compare the costs and benefits of a proposal, both measured by what people would be willing to pay. (As we'll see later, another way of applying the test would look at what people would demand in exchange for giving up a benefit, rather than what they would pay to acquire it. For now, however, we can ignore this complication.) Let's suppose we wanted to do a cost-benefit analysis of the land disposal remedy in *Reserve Mining.* On the benefit side of the equation, we would need to assign values, based on willingness to pay, for the elimination of this source of pollution. Some of these values would be relatively easy to determine, given the necessary scientific information—for example, loss in profits to fishing boats in Lake Superior caused by pollution. (At least the calculations would be straightforward in theory, although in trials it's not uncommon to find disagreements among expert witnesses relating to these lost-profit issues.)

Some other kinds of benefits are fairly easy to assess, at least in principle, because we can look at the choices people actually make in order to figure out their preferences. Examples of these measurable benefits include the recreational benefits enjoyed by tourists, which can be indirectly measured through the amount of time they're willing to invest traveling to lakes, and the quality-of-life benefits to local residents, which can be indirectly measured through the effect of lake access on house prices.[10]

Although we may start to feel some qualms at this point, we can even put dollar amounts on the health effects of the asbestos (provided, of course, that we can figure out just what those health effects may be). Without worrying right now about the technicalities, we could assign monetary values to different levels of risk by looking at how much consumers are willing to pay for safer products, or at how much income workers are willing to give up for safer jobs, or at how much travel time people are willing to sacrifice for the safety benefits of driving more slowly.

10. For a discussion of some of these methods, see David Pearce and R. Kerry Turner, *Economics of Natural Resources and the Environment* 140–148 (1990).

All of these would be different ways of determining the market value of safety.[11]

If people demand $1,000 in return for being exposed to a one in a thousand risk of death, it's conventional to say that the "value of life" is $1 million. This is a bit misleading, since they probably wouldn't be willing to commit suicide for that amount of money! To express this distinction, economists often speak of the value of a statistical (as opposed to individual) life.

We enter a different realm of difficulty with some other environmental benefits: so-called option and existence values. An example of option value might be posed by someone like me—although I don't have any particular plans to go to Lake Superior, I'd be willing to pay something in order to keep open the option of seeing the lake again if I choose to do so. Existence values are even more ethereal—for example, the amount of money I would be willing to pay to save rain forests, although my allergies make it extremely unlikely that I would ever go there. The benefits we discussed previously involved "use values," which actually flow from some direct physical interaction with a natural resources like Lake Superior. In contrast, nonuse values don't involve any observable current conduct.[12]

Nonuse values pose something of a dilemma for economists. Unlike use values, they can't be measured by looking at actual behavior in order to gauge preferences. But they clearly are real—many people feel very strongly about environmental issues such as saving whales or rain forests despite zero personal contact with the resource. Suppose it turned out that we could drain Lake Superior, sell the water to Los Angeles, and give everyone who lives, works, or plays near the lake enough money to

11. Of course, like many things that seem simple in principle, these measurements turn out to be rather difficult in practice, which is why economists have to spend so much time studying mathematical modeling and the like. The result is that, even if you buy the basic theory, the numbers can be quite soft. Although I'm putting aside this practicality for now, we'll see in the next chapter that it's extremely important.

12. For introductions to these concepts, see Christopher Stone, "What to Do about Biodiversity?: Property Rights, Public Goods, and the Earth's Biological Riches," 68 *S. Cal. L. Rev.* 577, 580–588 (1995); Timothy Swanson, "Economics of a Biodiversity Convention," 21 *Ambrio* 250 (May 1992).

make them happy with the change. Putting aside any question of whether this violates higher, noneconomic values, it isn't a Pareto improvement because many people with no direct contact with the lake would be appalled and refuse to consent. It may not even be economically efficient if the benefits to Los Angeles don't equal the amount that these bystanders would demand in order to go along with the change. Either the cost-benefit analysis must include these nonuse values, or the analyst takes the risk of misapplying the theoretical efficiency criterion, which supposedly justifies use of cost-benefit analysis in the first place.

As a solution, some economists advocate the use of "contingent valuation" studies to measure how much people are willing to pay for nonuse values. Contingent valuation is essentially a survey technique. People are given information about an environmental issue and then asked if they would be willing to pay a certain amount to solve the problem. There is a great deal of dispute about whether contingent valuation, even if done carefully, provides a genuine measure of preferences. Sunstein, for example, finds many contingent valuation analyses difficult to take seriously. He stresses what he describes as the "astonishing and devastating fact" that willingness to pay seems constant regardless of the scale of the environmental problem. In responding to surveys, he contends, "people may be purchasing moral satisfaction rather than stating their real valuation," merely proclaiming their unwillingness to feel responsible for environmental harms.[13] Economists critical of contingent valuation view the resulting numbers as mostly reflecting the warm glow that people get by announcing their support for the environment. These critics doubt that people actually have preferences about specific environmental sites or that their responses reflect considered efforts to assess such preferences.[14] But this view is by

13. Cass Sunstein, *Free Markets and Social Justice* 142–143 (1997). For an environmentalist critique of contingent valuation, see John Heyde, "Is Contingent Valuation Worth the Trouble?" 62 *U. Chi. L. Rev.* 331 (1995).

14. See Peter Diamond and Jerry Hausman, "Contingent Valuation: Is Some Number Better than No Number?" *J. Econ. Persp.*, Fall 1994, at 45, 56, 63; Brian Binger et al., "Contingent Valuation Methodology in the Natural Resource Damage Regulatory Process: Choice Theory and the Embedding Phenomenon," 35 *Nat. Resources J.* 443 (1995).

no means universal among economists. Advocates of contingent valuation argue that the critics have exaggerated the problems, that many problems can be limited through careful survey design, and that contingent valuation can be validated against other measures of environmental benefits.[15]

If anything, the trend may be running in favor of contingent valuation. Besides receiving a guarded endorsement from the federal courts,[16] a distinguished panel of economists advising the government has also given contingent valuation qualified support. Hard-nosed environmental economists like Paul Portney at Resources for the Future find it at least promising enough to deserve serious consideration.[17] The Clinton administration's guidelines for cost-benefit analysis allow agencies to use contingent valuation. The guidelines caution, however, that the state of the art is rapidly evolving. They also note that uncertainties about the technique argue for "great care in the design and execution of surveys, rigorous analysis of the results, and a full char-

15. See W. Michael Hanemann, "Valuing the Environment through Contingent Valuation," *J. Econ. Persp.*, Fall 1994, at 19, 21–26, 29–32; Katharine Baker, "Consorting with Forests: Rethinking Our Relationship to Natural Resources and How We Should Value Their Loss," 22 *Ecology L. Q.* 677, 714–720 (1995).

16. This seemingly esoteric academic dispute has legal significance because courts need to assess damages for harm to natural resources. Two federal statutes, the federal Superfund law (Comprehensive Environmental Response, Compensation, and Liability Act (CERCLA), 42 U.S.C. § 9607(1)(4)(c) (1998)) and the 1990 Oil Pollution Act (33 U.S.C. § 2706(d)(1) (1996)), require government agencies to establish rules for assessing natural resources damages. See Douglas Williams, "Valuing Natural Environments: Compensation, Market Norms, and the Idea of Public Goods," 27 *Conn. L. Rev.* 365 (1995). Under both statutes, a major question is whether contingent valuation can be used in assessing those damages. The use of contingent valuation in Superfund cases was upheld over an industry challenge in *Ohio v. United States Dep't of Interior,* 880 F.2d 432 (D.C. Cir. 1989). The court found that the agency's decision to adopt contingent valuation was made "intelligently and cautiously" and upheld adoption of the methodology as "reasonable and consistent with congressional intent, and therefore worthy of deference." Id. at 476–477. For further discussion of the litigation, see Judith Robinson, "Note: The Role of Nonuse Values in Natural Resource Damages: Past, Present, and Future," 75 *Tex. L. Rev.* 189 (1996). See also *General Electric Co. v. United States Dep't of Commerce*, 128 F.3d 767 (D.C. Cir. 1997) (upholding final contingent valuation rule).

17. See Paul Portney, "The Contingent Valuation Debate: Why Economists Care," *J. Econ. Persp.*, Fall 1994, at 3, 8–10, 15–16. The advisory panel was chaired by Nobel laureates Kenneth Arrow and Robert Solow.

acterization of the uncertainties in the estimates to meet best practices in the use of this method."[18]

Economists being the clever people that they are, I suspect that the methodological problems with contingent valuation will be solved—that is, that the results will be reasonably robust, internally consistent, and comparable to other valuations when those are available.[19] One particularly useful technique seems to be asking people how they would vote in a referendum to save a natural resource if the result was a specified increase in their own taxes, after providing them complete directions and background information.[20]

Assuming this methodology does eventually win acceptance by economists, the larger question is whether the results actually should play a role in making environmental policy decisions. Cost-benefit analysis treats environmental values as being on the same plane as ordinary consumer decisions: people are willing to pay to save the whales just as they are willing to pay for pizza delivery, except that it's more difficult to figure out the demand curve for whales. When a contingent valuation study asks people whether they would be willing to pay a higher tax in order to save a lake, they are basically being asked how much consumption of other goods they would give up to satisfy their "taste" for the lake. These are the kinds of preferences that are exhibited in markets, real or hypothetical. Since there is no functioning market for the existence value of whale species or lakes, the economist is forced to construct a hypothetical market, but it is still market conduct that provides the ultimate standard.

Private Lives versus Public Values

Now that we've gotten a better grasp on the concept of economic efficiency, it is time to consider the arguments against its use. Many legal scholars contend market behavior is simply the

18. Regulatory Working Group, *Economic Analysis of Federal Regulations under Executive Order 12866* (Jan. 11, 1996) <http://www.whitehouse.gov/WH/EOP/OMB/html/miscdoc/riaguide.html>.

19. On the other hand, properly conducted contingent valuations may turn out to be quite expensive. See Heyde, *supra* note 13, at 347 n.93.

20. See Hanemann, *supra* note 15, at 22–24.

wrong place to look for a basis for environmental policies. As Sunstein puts it, "[D]emocratic choices should reflect a process of reason-giving, in which it is asked what policies are best to pursue, rather than a process of preference satisfaction, in which each person is asked how much he is willing to pay for a certain result. Deliberative outcomes should not be confused with aggregated willingness to pay."[21]

Sunstein develops this point more fully as part of a critique of free market environmentalism, which he attacks for holding that "the market paradigm should be deemed normative for purposes of environmental protection." In Sunstein's view, "it is wrong to take private choices, expressed in the market domain, as definitional of preferences" because "[p]rivate willingness to pay in the market domain reflects a particular setting; it does not reflect global choices or valuations." In short, Sunstein maintains, "[t]he appropriate kind and degree of environmental protection raises issues that should be discussed by citizens offering reasons for one or another view," not determined by the choices made in their capacity as consumers.[22] Presumably, Judge Lord, if he were more philosophically inclined, would agree.

If this argument is correct, an official faced with an environmental policy choice should waste little time considering contingent valuation or other economic measures of environmental values. Instead, she should look to the domain of politics for assessments of environmental values.

The idea that environmental values occupy a higher plane than mere preferences is not new with Sunstein. In the legal literature, the first proponent was Mark Sagoff, a prominent environmental philosopher at the University of Maryland. Sagoff first became known to legal scholars for his role in the "plastic trees" debate.[23] In response to a suggestion in the journal *Science*

21. Sunstein, *Free Markets, supra* note 13, at 310–311.
22. Sunstein, "Legislative Foreword," *supra* note 5, at 304.
23. The major contributions to the debate were Laurence H. Tribe, "From Environmental Foundations to Constitutional Structures: Learning from Nature's Future," 84 *Yale L.J.* 545 (1975); Laurence H. Tribe, "Ways Not to Think about Plastic Trees: New Foundations for Environmental Law," 83 *Yale L.J.* 1315 (1974); and Mark Sagoff, "On Preserving the Natural Environment," 84 *Yale L.J.* 205 (1974). Tribe and Sagoff agreed in rejecting utilitarianism, but disagreed sharply in their rationales. Tribe argued that environmental values are

that plastic trees would be cheaper and more durable than real ones for some uses, he undertook to explain what was wrong with such a utilitarian calculation. An influential later book, *The Economy of the Earth*, is dedicated to attacking cost-benefit analysis and the associated standard of economic efficiency.

Sagoff's basic argument is that cost-benefit analysis applies only to personal preferences, whereas environmentalism involves moral values. Such moral issues, he believes, properly belong to the political process and cannot be resolved through economic analysis. Cost-benefit analysis can compare the competing weight of arbitrary private wishes, but not the weight of the moral principles involved in modern social regulation.[24]

The heart of this argument is a stark division between the interests people have as private consumers and those they have as citizens.[25] Public values are described as quite distinct from private preferences. Sagoff's thesis "is that social regulation expresses what we believe, what we are, what we stand for as a nation, not simply what we wish to buy as individuals."[26] Similarly, Frank Michelman, a leading legal theorist, says we should embrace the view that someone would vote for the Endangered Species Act even though she had no interest in environmental matters in her private life.[27] Sunstein agrees with the view that such statutes express cultural aspirations.[28] For example, he says, we may protect endangered species "partly because [society]

---

not "smoothly interchangeable" with economic values; hence, economic analysis is inappropriate. Sagoff agreed that environmental values differ radically from economic values, although his explanation was somewhat different. Because they are cultural or political rather than merely personal, Sagoff contended, environmental values transcend economics.

24. Mark Sagoff, *The Economy of the Earth: Philosophy, Law, and the Environment* 8–9, 17–18, 26–27, 113 (1988). He continues to hold the same view today about the irrelevance of preferences. See Mark Sagoff, "Muddle or Muddle Through? Takings Jurisprudence Meets the Endangered Species Act," 38 *Wm. & Mary L. Rev.* 825, 980–985 (1997). For a recent critique of the neorepublican viewpoint, see Douglas Williams, "Environmental Law and Democratic Legitimacy," 4 *Duke Envtl. L. & Pol'y F.* 1, 19–31 (1994).

25. See Sagoff, *The Economy of the Earth, supra* note 24, at 27, 50–51, 69, 94, 171.

26. Id. at 16–17.

27. See Frank Michelman, "Politics and Values or What's Really Wrong with Rationality Review?" 13 *Creighton L. Rev.* 487, 509–510 (1979).

28. Sunstein, *Free Markets, supra* note 13, at 369.

believes that the protection makes the best sense of its self-understanding. . . ."[29] Or, as Sagoff puts it, "Social regulation reflects public values we choose collectively, and these may conflict with wants and interests we pursue privately."[30] In somewhat different language, these scholars are communicating the same basic message about the chasm between public values and private preferences.

To see how this argument applies in a concrete context, consider the situation of the average citizen of Silver Bay, the town where Reserve had its operations. Shutting the company down would have had a devastating economic impact on the community's businesses and workers. Most of the residents would probably not have been willing to make this trade-off in private life. After all, if they preferred unemployment or bankruptcy to the health risks created by Reserve, they would always have the option of moving away. Nevertheless, Sagoff, Sunstein, and Michelman seemingly base their arguments squarely on the possibility that the residents would vote in favor of a shutdown in a referendum, putting the public interest ahead of their welfare. We are being asked to imagine Silver Bay residents acting entirely differently as voters than they do as workers, business owners, and consumers; their private decisions are supposed to be irrelevant to what they truly value as citizens. Such behavior seems far-fetched. Although we would like to believe that voters may take into account something beyond their own self-interest, the question remains whether positing a radical separation between personal preferences and public values is realistic (or desirable).

Of course, people's behavior does have contextual elements. All of us have our inconsistencies. But this dichotomy between public and private goes too far in splitting people's goals in two. Neo-republicans come close to denying any coherence to a person's character. Sunstein has most clearly expressed this view of human character:

> If someone takes a job that includes a certain danger, or chooses not to recycle on a Tuesday in March, or discriminates against a certain female job-candidate, we

29. Id. at 92.
30. Sagoff, *The Economy of the Earth, supra* note 24, at 17.

cannot infer a great deal about what he "prefers" or "values." All of these choices might be different in a different context. Our discriminator may support a law banning discrimination; people who do not recycle in March may recycle in May or June, and they may well support laws that mandate recycling.

He adds that economists believe they can derive general accounts of people's values from particular choices, a mistake that he says involves "extravagant inferences from modest findings."[31]

Obviously, people's choices are not always consistent, and they may not always live up to their aspirations. But surely there is something wrong with a picture in which private behavior and publicly espoused values are expected to have *no* connection. As one critic of Sunstein has said, it seems extraordinary to try to justify environmental law by invoking "the rather schizophrenic character who 'as an individual' loves 'sitcoms' but 'as a citizen' prefers Brahms."[32] Not to mention the sexist employer who unceasingly harasses female employees but is a sincere and staunch feminist "as a citizen." Assuming these eccentric (if not hypocritical) characters exist, surely we should not take them as our models.

The willingness of neo-republicans to ignore private preferences is fueled by a disdain for the value of private life. Sagoff, for instance, makes no secret of his dim view of ordinary life. "The pursuit of private satisfactions, except for the committed hedonist, soon becomes disappointing or boring, and we look for some public cause, like saving the whales, that does not benefit us personally but appeals to our conscience."[33] Sagoff's identification of hedonism and private satisfactions is telling. But soft drinks and television cartoons, while convenient targets,[34] are

31. Sunstein, *Free Markets, supra* note 13, at 6.

32. Stephen Williams, "Background Norms in the Regulatory State," 58 *U. Chi. L. Rev.* 419, 427–428 (1991) (reviewing Cass Sunstein, *After the Rights Revolution: Reconceiving the Regulatory State* (1990)).

33. Sagoff, *The Economy of the Earth, supra* note 24, at 116. His earlier writings were harsher: "Beauty . . . is valued as a source of pleasure. But pleasure is merely contemptible." Sagoff, "On Preserving the Natural Environment," *supra* note 23, at 209 (citing flattery and prostitution as sources of pleasure).

34. See Donald Hornstein, "Reclaiming Environmental Law: A Normative Critique of Comparative Risk Analysis," 92 *Colum. L. Rev.* 562, 625 (1992).

not the sum and substance of private life. Apparently, if we are to believe the neo-republicans, only a mindless consumer would derive satisfaction from reading novels, supporting a family, practicing medicine, or raising children.

This casual dismissal of private life is at best thoughtless, at worst a denial of important moral values. Indeed, Sagoff makes it clear that even minimal morality is not to be expected in private life. He tells us, with no evident embarrassment, that he bribed a judge to fix some traffic tickets, but then at election time helped to vote the corrupt judge out of office.[35] He speeds, but wants the police to enforce laws against speeding.[36]

Besides viewing private preferences as morally valueless, Sagoff seems to believe that the government can readily modify social values.[37] Similarly, Sunstein says that "private willingness to pay is undergirded by social norms and existing habits, and these should probably be changed; indeed, in the environmental context they had better be. . . . A prime purpose of environmental law is to shape norms and habits."[38] There is certainly reason to doubt the government's ability to accomplish this task proficiently. Despite decades of effort, the Soviet government was unable to motivate workers on collective farms to become productive. Equally stringent efforts to suppress religion were also notably unsuccessful. If totalitarian dictatorships cannot readily remold preferences, despite having complete control over the media and the educational system, it seems unlikely that democracies will have greater success. Although Sunstein is right to point out that preferences can sometimes change sharply, he provides no evidence of government success in deliberately fostering such preference shifts. Often the changes seem to happen spontaneously, with the government lagging behind, as in the environmental movement of the late 1960s.

As we have seen, the thrust of the neo-republican argument is to privilege the public over the private, drawing a strict line be-

35. See Sagoff, *The Economy of the Earth, supra* note 24, at 52.
36. Id. He also recounted that he had an "Ecology Now" sticker on a car that drips oil everywhere. Id. at 53.
37. See id. at 63–64, 93, 115–116. For a more guarded statement of this position, see Mark Kelman, *A Guide to Critical Legal Studies* 131–137 (1988).
38. Sunstein, "Legislative Foreword," *supra* note 5, at 304–305.

tween consumers and voters. The political sphere is elevated by viewing politics as uniquely deliberative, as shaping private preferences, and in general as having a higher moral status. Other scholars seem to make the reverse mistake by giving the privileged place to markets. This equally one-sided view privileges the market as an authentic expression of preferences, which are distorted by the corruptions of the political process. Reversing the romanticism of the neo-republicans, they hold politics in contempt.

This opposing view is sometimes found in public choice theory. Public choice theory is the application of economic methodology to political behavior. The public choice vision of politics can be as jaundiced as the neo-republicans' view of markets. One strand of public choice theory suggests that the political process is hopelessly dominated by special interests out to enrich themselves. For instance, public choice theorists do not paint a pretty picture of environmental legislation.

A thoughtful overview of the field is provided in a recent book by Jerry Mashaw, an administrative law scholar at Yale. As Mashaw says, "Rather than seeing the newly empowered consumers' or environmentalists' organizations as representatives of the public interest, critics have seen them as just new interest groups pursuing their own special ends." Many studies have "purported to show that the regulatory efforts of the new agencies [like the EPA] have produced modest improvements in the general welfare, while making massive redistributions of income from one group to another." Other studies "describe pernicious coalitions of traditional 'special interest' and newer 'public interest' groups which pursue converging organizational aims at the expense of both the public health and national economic growth." For instance, "Appalachian soft coal interests joined with Western environmentalists to produce air quality regulation that dirties the air unnecessarily at enormous economic cost."[39]

This strand of public choice suggests that politics is too corrupt to deserve our respect. Another strand of public choice theory holds that the political process is basically fortuitous, an

---

39. Jerry Mashaw, *Greed, Chaos, and Governance: Using Public Choice to Improve Public Law* 23 (1997).

arena ruled by chaotic voting patterns or arbitrary agenda set-
ters. If this vision is correct, the last thing we should do is trust
the political process as a guide to the public interest. Far better
to look to Main Street and Wall Street, where the venality is re-
strained by competitive markets.

To my mind, this completely jaundiced view of politics is as
unbalanced as the denigration of markets by Sagoff and others.[40]
Indeed, the two views refute each other. Governments and markets are both flawed, but useful, institutions. Our society is an
indivisible combination of public and private. It ill-behooves us
to fracture our social lives in half, picking one institution to cele-
brate as the fount of all virtue and maligning the other as the
root of all evil. If we are to evolve a vision of environmental law
fit for our society, we must recognize that we are deeply committed both to free markets and to democracy, as flawed as each
may be on occasion. We need to find a way to draw on both sides
in formulating environmental policy.

Only by acknowledging the claims of both the public and the
private spheres can we hope to create a durable scheme of environmental protection. Without appealing to public values, environmental regulations could not long enjoy general support
based purely on the calculus of competing private interests. But
without recognizing private interests as legitimate, environmental regulations may provoke unmanageable resistance from those
paying the price and are likely to be seen by society as a whole
as too draconian to be acceptable. Long-term, sustainable environmental regulations must appeal to public values, while recognizing the significance of economic interests as well.

Wedding Economics with Democracy

Beyond Reductionism

Economists do recognize a distinction between ordinary consumer preferences and environmental values, though not the
ethical distinction posited by their critics. For the economist, the
main difference is that environmental quality is a public good.

40. See Daniel Farber and Philip Frickey, *Law and Public Choice: A Critical
Introduction* (1988); Daniel A. Farber, "Politics and Procedure in Environmental
Law," 8 *J.L. Econ. & Org.* 59 (1992).

Briefly, a public good is a benefit that cannot be efficiently supplied by markets because there is no effective way to charge each "consumer" for the benefits received. (For example, if we improve air quality, we have no way of requiring payment as a condition for breathing cleaner air, so no company would have any way to market the benefit. Once we decide how much clean air to have, that is the amount available to be enjoyed by everyone.) Hence, public goods must generally be provided by the government rather than the market, not because they have a higher ethical value, but because of the practicalities of producing and distributing the benefits involved.

This economic perspective does provide some support for the neo-republican view of environmental law. Although the neo-republican position is badly overstated, it does provide some useful insights. Public deliberation is, as the neo-republicans maintain, crucial to environmental law. Market choices are usually decisions we make individually, without deferring to the views of outsiders. Thus, we are entitled to cast our "votes" in the marketplace without consulting anyone else. "Let them make their own choices," we might say. But because environmental quality is a public good, how each of us votes regarding the appropriate level of environmental quality affects everyone else. It is only fair, then, that we consider the views of others, rather than merely expressing our own personal preferences, because our votes can impose outcomes on others. To this extent, politics is properly more deliberative than markets, and this sort of deliberation is particularly appropriate for issues of environmental quality.

Although their argument is overblown, the neo-republicans also have a valid critique of economic reductionism, which views environmental values as merely inputs into a formal cost-benefit analysis. Cost-benefit analysis is based on a picture of human behavior that is often useful—a vision of rational actors seeking to maximize well-ordered, but otherwise arbitrary, preferences. But as my high school English teacher was fond of saying, "[T]he map is not the territory." The map is what you fold in your glovebox; the territory is what speeds by your car window. Likewise, the economist's vision is not definitive of human life, but only a useful picture. It would be a mistake to take a concept derived from the picture, like economic efficiency, and think that it is somehow the be-all and end-all of human values. Reducing all values to monetary terms would mean an unacceptable flattening of our normative landscape.

If it is wrong to view environmental values as merely inputs into a formal cost-benefit analysis, then we might well wonder whether economic measurements of value should play *any* role in decisions. Should we pay any attention at all to purported measures of the "value of life" or of the "contingent valuation" of endangered species and other aspects of nature?

For obvious reasons, many people find it repellent to put a dollar value on human lives or biodiversity. Nevertheless, there are two reasons why the economist's measurements should play a part in decision making. First, we should not lightly override decisions that workers and members of communities make about how to value levels of harm to themselves. After all, it is their lives, not ours, that are at stake. Second, we are badly in need of benchmarks to assess the general reasonableness of environmental decisions. As I have argued above, environmental values should not be wholly immune from comparison with economic preferences. The economist's techniques for measuring environmental benefits offer one way of making that comparison. Economic measurements should be taken not as gospel, but as useful starting points.

The Perils of Paternalism

One advantage of the economic efficiency standard is that it calls our attention to the issue of paternalism in regulating health risks. We can illustrate the paternalism problem by considering the risks to workers in *Reserve Mining*, putting aside for the moment the impact on the larger community. Reserve employed about three thousand workers. Suppose current asbestos exposures (including air and water) will eventually kill three of these workers. Eliminating the asbestos will cost $240 million, which comes to $80 million per life saved.[41] If the employees do not

---

41. These figures are actually the same order of magnitude as those discussed in an important Supreme Court case, which is considered in chapter 3. See *Industrial Union Dep't v. American Petroleum Inst.*, 448 U.S. 607, 628–630, 654 (1980) (plurality opinion).

value their own lives that highly, a cost-benefit analysis will be unfavorable.

This unfavorable cost-benefit analysis means that in some sense we are making society as a whole poorer. For instance, suppose workers actually value their lives at $40 million apiece (which is much higher than the amounts revealed by econometric studies). Then the employer is spending $240 million to deliver a benefit that the employees assess as being worth $120 million to them.[42] Given these figures, neither labor nor management would support government regulation. Both sides would prefer that the government simply tax the company $180 million and send a $60,000 check to each worker. From their perspective, banning the toxin is no different from imposing a $240 million tax on the company, burning half of the money, and giving the rest to the workers. Some workers may even find themselves jobless as a result of the regulation.[43] Even those who keep their jobs may well pay indirectly for the safety benefits.[44] Thus, the regulation apparently is worthwhile only if we believe that the workers misunderstand the facts or undervalue safety. The ultimate issue, then, is paternalism.

Sagoff has little hesitation in embracing paternalistic regulation: "The transactions that led to child labor, the sixteen-hour workday, and hideous workplace conditions were largely voluntary and informed; no centralized bureaucracy in Washington

42. Taking into account risk aversion would complicate, but not fundamentally change, the analysis.

43. For example, suppose safety costs are proportional to output. Then the effect of the safety regulation is to shift the firm's supply curve to the left. (It takes a higher price to induce the firm to produce the same quantity.) The result is that output falls, as does employment.

If the firm has market power in the relevant industry, then as a result of the decreased supply, the product price also rises. Hence, part of the safety cost is passed on to consumers in the form of higher prices. This is not altogether good news for the workers because some of them still lose their jobs.

44. Their relative bargaining position limits the total compensation that they can extract from the employer. The government cannot increase their total compensation by ordering that part of it be paid in the form of safety improvements any more than it could do so by ordering that workers be paid with apples instead of dollars. If workers are unwilling to buy apples or safety at the market rate, it is hard to see how their economic well-being can be increased by requiring employers to pay part of their salary in those forms.

told workers how old they had to be or what minimum they had to be paid; labor markets were efficient." But "the resulting levels of death, misery, and disease, even if 'optimal' or 'efficient' from an economic point of view, cannot be tolerated in any civilized nation."[45]

The reason Sagoff does not find paternalism even remotely troubling is that he is convinced that safety regulations upgrade workers' preferences: "A hundred years of compassionate legislation has produced conditions in which economists can now argue that voluntary markets set an appropriate value on worker safety. This is a result not of more efficient markets but of persistent ethical regulation."[46] This argument attempts to avoid the paternalism problem by relying on changing preferences. Because of the regulations, workers will become accustomed to greater safety, and their willingness to tolerate risk will decrease. Hence, after the regulations are in effect long enough, they will fit the workers' own preferences. So, Sagoff seems to think, paternalism is not a problem.[47]

This argument is deeply flawed.[48] Paternalism may sometimes be justified. Sometimes people don't seem capable of making decisions that are clearly in their own best interests. But we should be wary of jumping too quickly to the conclusion that we know their interests better than they do. This may merely be an excuse for imposing our values on them, the domestic equivalent of environmental imperialism. For example, the workers in *Reserve Mining* were clearly exposed to health risks as the result of the asbestos problem. Just as clearly, they were vehemently opposed

45. Sagoff, *The Economy of the Earth*, *supra* note 24, at 15.

46. Id. at 116.

47. See id. at 16, 115–116.

48. This argument has two flaws. First, it may not be factually accurate. How do we know that workers will be more willing to sacrifice wages for safety after the regulation is in effect? Maybe, despite the regulations, they will still value the lost wages more than the increased safety. Second, as a general matter, reliance on future preference shifts is a dangerous way of avoiding the paternalism problem because it has the capacity to justify almost anything. See David Shapiro, "Courts, Legislatures, and Paternalism," 74 *Va. L. Rev.* 519, 549–550 (1988) (rejecting this argument for paternalism); Cass Sunstein, "Legal Interference with Private Preferences," 53 *U. Chi. L. Rev.* 1129, 1148–1150 (1986) (endorsing this form of paternalism, but cautioning against the risk of abuse).

to closing the plant, with the resulting devastating impact on local communities. It might be easy enough for us to say, at a comfortable distance, that a plant closing would have been the right choice because health always comes ahead of mere economics. But the workers might well have viewed this as manifesting not compassion, but a callous disregard of their values and interests. Sometimes we may feel that we have no choice but to intervene in such ways, hoping that our values will turn out in the end to be vindicated. But we ought to look carefully at the preferences of the people most involved before we intervene.

It is true, as Sunstein and others have argued, that workers' decisions about risk are contextual and potentially distorted. For this reason, they should not automatically be accepted as valid indicators. But voters' decisions about risks may also be contextual and potentially distorted, rather than representing true preferences. While it is true that workers may have insufficient information, limited options, and difficulty in processing risk data, the same is true of voters. Indeed, voters have probably less reason to inform themselves about any particular issue than workers do about their workplaces, their options may be even more limited because of the constraints of the two-party system, and sitting in a voting booth does nothing to enhance a person's ability to perform a statistical risk analysis. Moreover, since voters typically select candidates rather than voting directly on issues, we can only reconstruct the values underlying their choices with some difficulty. Yet in a democracy, there are limits to how well the government can outperform the voters in defining public policy. Thus, the possible flaws in individual decision making may argue as much against trusting the government as against trusting the market. Cost-benefit analysis at least alerts us to situations where we need to think seriously about the paternalism issue.

The Relevance of Economic Measurements

Willingness to pay also has the advantage of requiring voters to "play with their own money." Perhaps voters are willing to vote for higher levels of safety than they are willing to purchase in private life because of public-regarding social values. On the other hand, perhaps they may be willing to vote for higher levels of safety because they believe that someone else will be paying for them. "Let's spend *their* money to achieve *our* goals" is an inherently attractive proposition. Decisions of this kind may show how strongly voters feel about their goals or merely how little they mind spending their neighbors' money. "Willingness to pay" avoids this issue.

Thus, one reason to pay attention to willingness to pay is that it may sometimes convey information that isn't captured by willingness to vote. Indeed, sometimes we may be otherwise at a loss even to begin to assess environmental values. Suppose the only ill effect of the taconite dumping was the green water, and also suppose this had no significant effect on the Lake Superior ecology. Still, one of the wonders of Lake Superior is the clarity of its waters, and it should be worth something to preserve that clarity as an aesthetic value, just as we are willing to spend money to preserve visibility at the Grand Canyon. But what is a reasonable amount to spend for this purpose? Ten million dollars? A hundred million?

In thinking about this issue, one helpful starting point is to get a sense of how strongly one's fellow citizens feel about the issue. Would the citizens of Minnesota and Wisconsin be willing to contribute $100 apiece in extra taxes to preserve the purity of the lake? A contingent valuation wouldn't provide the final answer to the question of how much to spend, but at least it might provide a useful starting point. Indeed, despite his general skepticism of contingent valuation, Sunstein seems to acknowledge this function.[49]

Thus, we should reject the two extreme positions. Bean counters are wrong to assume that market preferences define the public interest. But tree huggers are also wrong when they call market preferences irrelevant to public policy. They have a legitimate, but not defining, role to play.

It is tempting for those of us who consider ourselves environmentalists to reject economists' assessments of environmental values. These measurements are based on a reductionist theory that fails to do justice to the richness and depth of our normative

49. Sunstein, *Free Markets, supra* note 13, at 310.

commitments. Willingness to pay does not come close to capturing all environmental values, and it would be foolish to base environmental law solely on this standard. Markets are flawed as arenas in which people express their personal values. But, then, so is politics. Both offer a blurry and sometimes distorted view of our society's judgments. For this reason, we cannot afford wholly to ignore either one in making environmental policy decisions.

Finding the Public's Values

To try to see how all of this works out in more concrete terms, suppose a decision maker is trying to determine whether to allow a large wetland to be filled. The houses built on the wetland will be collectively worth $10 million more than their construction cost, and there is no other feasible location in the area. On the other hand, there is no other wetland nearby either. The city council does not have jurisdiction over the project, but has passed a resolution demanding that the wetland be saved. One survey shows that residents of the metropolitan area would be collectively willing to pay up to $15 million to save the wetland. Which pieces of information are relevant? At this point, I don't want to try to address the ultimate question of what the decision should be, or even the exact method of reaching a decision, but merely what factors are legitimately considered.

Let's start with the value placed on the homes, which is $10 million in excess of their construction costs. Does this figure deserve any weight in the analysis, or does it reflect mere private preferences with no public value? There may be distortions in the housing market due to government subsidies or anticompetitive conditions, but we can presume until proven otherwise that the $10 million really does measure what buyers would be willing to pay for the houses over and above the cost of construction. In short, they apparently believe that it is worth giving up that amount in other forms of consumption in order to have the houses. Unless we have some reason to doubt their ability to make such decisions—and in a capitalist society, we are more or less forced to assume the contrary—then we should defer to their judgment about the value of the homes to them.

Nor is there any reason to dismiss the houses as mere consumer baubles, unworthy of consideration by society as a whole. After all, homes are basic to people's way of life, and a society could hardly be commended for ignoring either the basic need for housing or the desirability of attractive, convenient dwellings. In short, what the market is telling us is that, based on the basic standard our society uses to make most decisions about what goods to produce, these houses are worth producing if we put to the side the value of the wetland.[50]

What about the $15 million value placed on the wetland in the survey? Whether or not this number represents what an economist would call a genuine preference, it does express how much members of the public say they value the existence of the wetland. Assuming that the polling was done carefully, we can reasonably say that this figure does express a collective view about the importance that the public places on the wetland. The city council's action reinforces this conclusion. Just as our acceptance of capitalism requires us to consider the values expressed in the market, our commitment to democracy requires us to give credence to collective expressions of value.

Still, one might ask whether this collective view is not itself distorted in some way. Economists who object to current environmental policy tend to think that voter support for an environmental policy is often misinformed or irrational. They argue that the government "should not mimic these shortcomings in individual behavior, but rather it should make the kinds of rational and balanced decisions that people would make if they could understand risk sensibly."[51] Surely, we don't want government policy to be based on irrational hysteria about environmental problems.

Before blindly accepting public values, as expressed in public opinion polls, legislation, or otherwise, we need to consider whether those views represent a considered, reasoned judgment

50. Of course, this isn't an irrebuttable judgment. Maybe there is evidence that the housing market is distorted in some way or that people systematically overestimate the value of housing because they fail to consider the full cost of their mortgage payments. But presumptively, the $10 million represents a genuine indicator of social value.

51. W. Kip Viscusi, "Regulating the Regulators," 63 U. Chi. L. Rev. 1423, 1447 (1996).

or an irrational impulse. If it is done properly, the poll may accurately express the public's values, but it cannot tell us whether those values are misguided. But the case for dismissing public environmental values as hysterical or misinformed has not been made, and we have some reasons for questioning it.

First, if environmental values were based on ill-considered prejudices or bias, we could make some predictions about how they would be distributed in the population. Presumably, irrational environmental values would tend to be strongest, all things being equal, in groups that are relatively ill informed or incapable of critical thinking. We would expect children to have less well considered views than adults and the poorly educated to be less well informed than others. So we would expect environmental values to be stronger among these groups than among well-educated adults. But the reality is to the contrary: environmental values become stronger and more sophisticated as children undergo intellectual development, and well-educated adults are markedly more pro-environmental.[52] Perhaps these phenomena have other explanations, but on their face, they fail to support the irrationality thesis.

Second, the public's views have at least some support in more hard-nosed analyses of the value of the environment. Although the numbers expressed in contingent valuation studies for the value of wetlands, coastal areas, and other environmental amenities sometimes seem high, they may have some foundation. A study recently published in the scientific journal *Nature* put the economic value of the "services" performed by the global environment at $33 trillion, and the authors considered this a conservative estimate. For example, the water purification services of the Catskill Mountains were assessed at $4 billion (the amount it would cost New York City to build a purification plant to achieve the same results).[53]

52. Stephen Kellert, *The Value of Life: Biological Diversity and Human Society* 46–51, 54–56 (1996).

53. See William Stevens, "How Much Is Nature Worth? For You, $33 Trillion," *N.Y. Times,* May 20, 1997, at C1. The original report is Robert Costanza et al., "The Value of the World's Ecosystem Services and Natural Capital," 387 *Nature* 253 (1997). For a more extensive discussion of this approach, see *Nature's Services: Societal Dependence on Natural Ecosystems* (Gretchen Daily ed., 1997).

In a democratic society, the values expressed in legislation should be treated as presumptively valid. Of course, in particular cases, legislation may be partly the result of public hysteria of the type discussed above or the outcome of special interest lobbying. (In chapter 6, I will suggest a mechanism for correcting some of these legislative misfires.) But the case has yet to be made that this is generally true for environmental regulation as a whole. In the meantime, it would be wrong to dismiss either the market or the political system as a source of information about the public interest. In Robinson Crusoe's world, economics and politics are identical, and both reduce to a question of his personal goals. In a world containing many individuals with conflicting goals, making a societal decision is inevitably complex. Politics and economics provide alternative mechanisms for expressing and reconciling these individual goals.

Where does all of this leave the decision maker for our hypothetical wetland project? What we have done, so far, is to tell her to consider a rather disparate set of factors: the economic value of the housing project, the contingent valuation, and the considered view of the community's democratic representatives. How to weave these factors together into a concrete decision is the hard question. In the next two chapters, I will argue that decision makers should begin with an environmental baseline, allowing environmental damage only when avoiding it is infeasible or grossly disproportionate in cost. Putting aside the question whether the developer should be compensated for the lost investment, it seems unlikely that the housing development would be justified under this test.

If we imagine a decision maker holding a hearing about the correct course of action, this chapter has in a sense concerned the rules of evidence that should apply at this hearing. The conclusion is that both consumer preferences and political choices are relevant evidence. This is not an insignificant point, since such strong arguments have been made to exclude one or the other form of evidence altogether. But once the evidence is admitted, the decision maker still faces the hard problem of how much weight to give each item in drawing a conclusion. So far, we have established only something that may seem obvious to some readers: that environmentalism (as expressed in a series of environmental statutes) and economics (as expressed in the market) both have something to tell us about public policy. In the next chapter, we will examine how the decision-making techniques associated with each view operate and what role each should play. Once again, the context will be the *Reserve Mining* case.

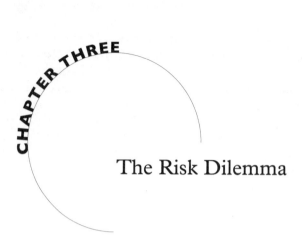

The Risk Dilemma

Economic and environmental values both have roles to play in the analysis. But what method should we use to factor them together? Much energy has been expended in a battle between advocates of two different methods, roughly corresponding to the tree huggers and bean counters of the preceding chapter.

To see how these methods might work, consider the problem of regulating kryptonite, a fictional pollutant. One regulatory method focuses on achieving the maximum feasible level of environmental quality. Once we determine that kryptonite poses an environmental threat, we would want to eliminate the threat to the extent possible. We might do this by requiring all polluters to use the best available technology (often called BAT) for controlling kryptonite emissions, or we might direct them to take all feasible steps to lower emissions to a safe level. I will refer to this as the feasibility approach. The other regulatory method is cost-benefit analysis, under which regulatory decisions are made by balancing the costs and benefits of regulation. The struggle between advocates of these two methods has consumed many a tree.

In my opinion, this debate has suffered from a certain unreality. Ultimately, the most important practical question is not the choice of one exclusive methodology. Rather, it is how best to use whatever tools are available to make intelligent judgments in hard cases. Regarding these issues, close attention to a concrete example can do a great deal to advance the analysis. *Reserve Mining* provides an excellent case study in the uses and shortcomings of both methodologies.

Some readers may be tempted to avoid both methodologies by eliminating any consideration of cost. As Judge Miles Lord ultimately did, they may find *Reserve Mining* an easy case on the grounds that human life should always outweigh mere cost, so that Reserve should have been closed immediately to eliminate any health risk. But *Reserve Mining* is actually an excellent illustration of why both cost and risk have to be considered. A shutdown would not merely have taken money out of the pockets of some wealthy shareholders. It would also have imposed enormous costs on the company's employees, their communities, and the economy of the whole region. Even the construction involved in switching to land disposal posed some risk to workers. Whether land disposal would save any lives was unclear. There is some possibility—in the view of the Eighth Circuit, at least a 50 percent chance—that eliminating asbestos from Lake Superior would not in fact save a single human life. And if we were looking for ways to save lives, perhaps we could find better ways to spend over $200 million, such as subsidizing vaccinations for the poor. So the mere abstract possibility that land disposal might *conceivably* save a life is simply not enough to justify the order against Reserve. Some consideration of cost seems inevitable.

Other readers may find it obvious that land disposal was a misguided environmentalist boondoggle. After all, there was no proven threat to health, and $200 million is a lot to spend to fend off a speculative possibility. But it would be a mistake to jump to this conclusion. Admittedly, $200 million is a lot of money, but the company's activities were placing at risk the lives of a hundred thousand residents of Duluth. Imagine that Reserve had continued its activities for another twenty years and that we later learned that it had caused several hundred additional cancer deaths. Now imagine instead that Reserve was forced to spend the money, but that later scientific developments proved the asbestos harmless. Which outcome would we regret the most after the fact? We would probably experience far more regret for a mistaken gamble with the city's public health than a bad gamble with the company's money. At least it's not at all obvious that the court made the wrong choice about this trade-off.

We need to take both the cost and the benefits of regulation into account in some way. This, of course, is easier said than done. Given the difficulty of assigning values to the costs and benefits of health regulations, regulatory decisions clearly cannot be made by any mechanical formula. It is tempting to fall back on sheer intuition. But intuition is a treacherous guide. When the decision is being made by an administrator or a judge, we would like to have a little more guidance than simply the decision maker's "gut reaction." Too many different kinds of people get jobs as administrators and judges for us simply to trust their unguided intuitions; that's why we don't simply authorize government officials to operate on unarticulated hunches. Certainly, there are reasons for trepidation in trusting our destinies to the untamed impulses of a Miles Lord or his conservative counterparts.

Even when we are making decisions ourselves, intuition may fail. In thinking about *Reserve Mining,* my own intuition has waffled. On some days, it seemed clear to me that even one extra case of cancer in Duluth would be one too many and that Reserve had no right to gamble with public health. On other days, the thought of spending $200 million to combat such a speculative risk seemed bizarrely unrealistic. Perhaps I am unusual in experiencing these conflicting impulses, but I doubt it. In reality, most people's feelings are a confused mixture of tree hugger and bean counter. Intuition may be indispensable, but by itself, it is not enough to give us confidence in our decisions.

As discussed above, two methods have been suggested, in place of raw intuition, for making risk decisions. One is a cost-benefit analysis, comparing the predicted body count and the cost of control. The other method calls for avoidance of risks whenever feasible. As Mark Sagoff has suggested, feasibility analysis appears to be based on the view that people have a right to be protected from environmental risks.[1] Feasibility analysis

---

1. Mark Sagoff, *The Economy of the Earth: Philosophy, Law, and the Environment* 185, 197–198 (1988); see also *Weyerhauser Co. v. Costle,* 590 F.2d 1011, 1043 (D.C. Cir. 1978) (noting that the best practicable technology standard under the Clean Water Act embodies the view that the public has a right to a clean

stresses nuances (such as voluntariness, strangeness of risk, and concentration of costs on particular firms), as opposed to overall cost and mortality reduction. The biggest difference, however, is that feasibility analysis incorporates a strong presumption in favor of control, while cost-benefit analysis rejects any such presumption and calls for balancing.

Advocates of both views agree on the need for cost effectiveness; whatever kind of risk reduction we decide on, we should try to achieve the goal as inexpensively as possible. The dispute is about how to set the goal. Should we think about environmental quality first and cost second, or should we give them equal weight? Before addressing those issues, we need to take a closer look at how both approaches work in practice.

Maximum Feasible Regulation?

When first thinking about toxics problems, many people begin with the notion that carcinogens are bad things and should be eliminated from the environment at all costs. Statutes written in this way are usually stymied in the implementation phase because society simply is unwilling to close down entire industries.[2]

Although it seems plausible (who is in favor of cancer?), this absolutist approach turns out to be simply untenable. We are surrounded by natural carcinogens such as chemicals naturally found in a variety of foods.[3] Even apart from these natural substances, our ability to measure tiny traces of chemicals makes the idea of complete purity obsolete.[4] Regulation can even be counterproductive. We may eliminate preservatives that cause a

environment, limited only by the extent to which cleanup is impractical or unachievable).

2. See, e.g., Frank Cross, *Environmentally Induced Cancer and the Law: Risks, Regulation, and Victim Compensation* 72–73, 104–107 (1989) (discussing the EPA's inability to implement hazardous emissions provisions of the Clean Air Act). See also Frank Cross et al., "Discernible Risk: A Proposed Standard for Significant Risk in Carcinogen Regulation,"43 *Admin. L. Rev.* 61, 63–65 (1991).

3. See Bruce Ames et al., "Ranking Possible Carcinogenic Hazards," 236 *Science* 271, 273 (table ranking possible carcinogenic hazards), 276–277 (discussing natural carcinogens).

4. See Cass Sunstein, *After the Rights Revolution: Reconceiving the Regulatory State* 88–89, 95 (1990) (discussing the adverse effects of the Delaney Clause, which prohibited the Food and Drug Administration (FDA) from permitting any food additive found to be carcinogenic).

small risk of cancer only to foster the growth of pathogens that will cause deadlier risks of food poisoning.[5] Similarly, unrealistic consumer safety regulations may increase the number of deaths by driving up prices to the point that people shift to more dangerous unregulated substitutes.[6] As two legal scholars poignantly put it, "Risk inheres in our condition. Whether brought on by nature in such forms as earthquakes and disease, or by humans with mundane machines like the automobile and high technologies like nuclear energy, hazard is ubiquitous and inevitable."[7]

The fallback position is to require only as much elimination of health risks as is "feasible." A typical mandate is found in the current version of section 112 of the federal Clean Air Act,[8] requiring use of the maximum available control technology for toxic air pollutants. In a variety of statutory formulations, this has become the dominant mode of regulation in contemporary environmental law.[9] In this section, we will explore this approach and its application to *Reserve Mining*.

In the so-called *Benzene Case*, the Supreme Court clarified how feasibility analysis is applied to toxic chemicals like asbestos. Before regulations can be imposed, the Court requires a threshold showing of a significant risk of harm at current levels of ex-

5. Id. at 89.

6. See Frank Cross, "The Perils of the Precautionary Principle," 53 *Wash. & Lee L. Rev.* 851, 867–882 (1996).

7. Clayton Gillette and James Krier, "Risk, Courts, and Agencies," 138 *U. Pa. L. Rev.* 1027, 1027–1028 (1990).

8. 42 U.S.C. § 7412 (1998).

9. See Cross, *Environmentally Induced Cancer and the Law, supra* note 2, at 90–93. See also Christopher Schroeder, "In the Regulation of Manmade Carcinogens, If Feasibility Analysis Is the Answer, What Is the Question?" 88 *Mich. L. Rev.* 1483, 1486 (1990). One form of feasibility analysis would be to require industries to remove "known carcinogens" as much as possible (as opposed to the absolutist demand to eliminate them entirely). Even this less absolutist approach poses difficulties. *Reserve Mining* illustrates one difficulty of such an approach: should we consider "asbestos" to be a known carcinogen because its inhalation causes cancer, or should we say that "ingested asbestos" is not a known carcinogen? Even for a known carcinogen, if current exposure levels present only a minute risk, should we still require the industry to spend huge amounts on pollution control?

posure to a toxic substance.[10] The case involved an Occupational Safety and Health Administration (OSHA) regulation governing exposure to benzene. The secretary of labor had found that benzene was a carcinogen and that no known safe level existed. Hence, he lowered the permissible exposure level for workers from 10 ppm (parts per million) to 1 ppm, which he considered the lowest feasible level. Note that OSHA's reasoning ignored any comparison of the costs and benefits of regulation.

Indeed, OSHA never made a serious effort to estimate the benefits, and the costs were substantial. OSHA estimated the initial cost of compliance at almost half a billion dollars, followed by $34 million in annual costs. About thirty-five thousand workers would benefit from the regulation. The industry claimed that the regulation was unjustified because, even under the most conservative method of estimating risk, current exposure levels would cause at most two deaths every six years. OSHA rejected this argument, perhaps partly because it doubted the industry's factual claim, but apparently also because it regarded the level of risk as irrelevant: "[D]ue to the fact that there is no safe level of exposure to benzene and that it is impossible to precisely quantify the anticipated benefits, OSHA must select the level of exposure which is most protective of exposed employees."[11]

It may be useful to pause and compare the facts of the *Benzene Case* with those of *Reserve Mining*. Both cases involved chemicals known to be carcinogens at higher exposure levels, but whose dangers at low levels were unclear. The exposed population in the *Benzene Case* was about a third the size of the population of Duluth, and the cost of compliance was over twice as high, so even if the risks were equivalent, the cost effectiveness of land disposal in *Reserve Mining* was actually about six times better than that of the benzene regulation. Also, in *Reserve Mining,* the risk impacted a single community, whereas in the *Benzene Case,*

10. *Industrial Union Dep't, AFL-CIO v. American Petroleum Inst.*, 448 U.S. 607, 642 (1979) (*Benzene Case*). The lead opinion was written by Justice Stevens and is the basis for the discussion in the text. Technically, it was only a plurality opinion. (Justice Rehnquist concurred for other reasons, while four Justices dissented, so it was a 4–1–4 decision.) Lower courts and commentators, however, have treated the Stevens opinion as defining present law.

11. Quoted in id. at 654.

the risk was spread among a diffuse group of workers. Finally, in the *Benzene Case*, unlike *Reserve*, there had been an actual attempt to quantify the risks, but the agency had shrugged off the results of the analysis.

The Supreme Court overturned the benzene regulation because of OSHA's failure to take the low estimate of risk more seriously. According to the Court, the burden was on OSHA to show, "on the basis of substantial evidence, that it is at least more likely than not that long-term exposure to 10 ppm of benzene presents a significant risk of material health impairment."[12] Although it reversed the agency, the Court did show some sympathy for the difficulty of risk regulation. The benzene opinion clearly allows the agency to make conservative policy judgments in estimating the risk, even in the face of considerable scientific uncertainty. Although the opinion has been criticized for putting too heavy a burden on the agency,[13] it does give the agency room to maneuver so long as the agency is explicit about defining the significant level of risk. The Court made it clear that the significant risk requirement is not a "mathematical straitjacket" and that it is up to the agency to determine what it considers to be a significant level of risk. Moreover, the agency has leeway in considering the evidence and is "not required to support its finding that a significant risk exists with anything approaching scientific certainty." Hence, "so long as they are supported by a body of reputable scientific thought, the agency is free to use conservative assumptions in interpreting the data with respect to carcinogens, risking error on the side of overprotection rather than underprotection."[14]

The Court had trouble explaining what it regarded as a significant level of risk. At one point, the Court said a one-in-a-thousand risk "might well" be considered significant. If "the odds are one in a thousand that regular inhalation of gasoline vapors that are two percent benzene will be fatal, a reasonable

12. Id. at 653.
13. See generally Howard Latin, "The 'Significance' of Toxic Health Risks: An Essay on Legal Decisionmaking under Uncertainty," 10 *Ecology L.Q.* 339 (1982).
14. 448 U.S. at 655–656 (plurality opinion). See also id. at 663 (Burger, C.J., concurring); id. at 666–667 (Powell, J., concurring).

person might well consider the risk significant and take appropriate steps to decrease or eliminate it."[15] This seems reasonable enough as applied to a single exposed individual, but may not be so reasonable when a large number of people are exposed. Consider a one-in-a-thousand risk applied to the American population at large. As applied to the general population, such a risk level would mean approximately 240,000 deaths. Not only "*might* a reasonable person consider such a risk significant," but only a highly unreasonable person would think otherwise! Even a substantially smaller risk, killing a tenth or even a hundredth as many people, might well be considered significant, given such a large population at risk.[16]

Another related question about the "significant risk" standard involves scientific uncertainty about the risk level. On one reading, the benzene opinion requires the agency to make a finding that a chemical "more likely than not" causes some increase in mortality or disease.[17] If taken literally, this could lead to absurd results. Given a 60 percent chance that a substance was harmless, the agency could never act—not even if there was also a 40 percent chance that the substance might kill a million people. The opinion is more plausibly read as saying that the agency must find a significant risk of harm, taking into account both the likelihood that there is *any* increase in mortality and the probable extent of the increase if there is one. So, even though the Eighth Circuit seemed to think there was probably no increase in mortality caused by the asbestos, it may still have been right to order a remedy in *Reserve Mining*. A rational decision should be based not purely on the estimate of the most likely level of risk, but also on the seriousness of the possibility that the actual harm would turn out to be much higher.

15. Id. at 655.
16. Moreover, we might very well be inclined to consider a risk more significant if the resulting deaths were clustered, so that whole families or communities were especially impacted, as opposed to being spread out evenly across the population. (We might be particularly concerned if the victims were clustered in groups that are otherwise disadvantaged anyway, like racial minorities.) In short, the concept of significant risk is multidimensional.
17. 448 U.S. at 653 (plurality opinion). See also Latin, *supra* note 13, at 344–349 (critiquing plurality's allocation of burden of proof).

In a later opinion interpreting the same statute, the Court held that, once a significant risk is found, the agency must assure workers' safety to the extent feasible. Cost-benefit analysis is not required or even appropriate. The Court relied on the dictionary meaning of "feasible" as "capable of being done, executed, or effected." Thus, the Court said, the statute requires the agency to "issue the standard that 'most adequately assures . . . that no employee will suffer material impairment of health,' limited only by the extent to which this is 'capable of being done.'" In short, the Court said: "Congress itself defined the basic relationship between costs and benefits, by placing the 'benefit' of worker health above all other considerations save those making attainment of this 'benefit' unachievable."[18]

From an economist's point of view, the benzene approach is both over- and underprotective. This approach is overprotective because if the agency finds a substantial risk, it must regulate to the hilt without keeping costs in proportion to benefits. But it is also underprotective because it ties the agency's hands when an "insignificant" risk could be eliminated very cheaply.

In applying the benzene approach to *Reserve Mining*, the first step would be to determine if there was a significant risk. Unfortunately, even today, we cannot be sure of the answer. The type of asbestos involved in the case (amphibole rather than chysolite) is currently considered by far the more dangerous.[19] There has been a great deal of research on the carcinogenic properties of ingested asbestos, but the results remain inconclusive. The increased rate of gastrointestinal cancer among asbestos workers is well established. The "coughing up and swallowing" (more politely known as pulmonary clearance) hypothesis remains well accepted. Also, at least one careful epidemiological study, from the San Francisco Bay area, did find a significant increase in cancer related to asbestos in the drinking water. Other studies (including a follow-up study of Duluth) point in the opposite direc-

18. *American Textile Mfrs. Inst. v. Donovan*, 452 U.S. 490, 509 (1981) (*Cotton Dust Case*).

19. See Lee Siegel, "Note: As the Asbestos Crumbles," 29 *Hofstra L. Rev.* 1139, 1145–1146, 1157–1163 (1992) (citing current medical literature).

tion.[20] Putting all this information together provides no obvious answer.

Although the risk evidence today may be less worrisome, it seems more reasonable to assess the court's decision in terms of expert opinion fairly close in time to the decision. Ideally, only information available at the precise time of the decision would be considered. To take advantage of the EPA's expertise, however, I have extended the relevant period a few years. Rather than making my own judgment about the medical literature, I have chosen to rely on a possibly biased (and not quite contemporaneous), but at least much more knowledgeable, assessment by the EPA.

The EPA made this quantitative risk analysis a few years after *Reserve Mining*. In 1983, the EPA conducted a careful survey of the asbestos research and issued its own assessment of risk.[21] A later article converted the EPA assessment into a formula for estimating mortality rates.[22] The EPA estimate was based on lifetime exposure of 1 million people to 100 million fibers per liter of water. The EPA estimated an increased risk of death of 1/300 per person, or a total of 3300 excess deaths per million.

Duluth's population was about one hundred thousand, and a reasonable estimate of average fiber density was about 33 million fibers per liter.[23] (This average level was about five times higher

20. Citations to the technical studies can be found in Daniel Farber, "Risk Regulation in Perspective: Reserve Mining Revisited," 21 *Envtl. L.* 1321, 1344 n.109 (1991).

21. Assuming ingestion of 2 liters of water per day over a seventy-year lifespan, the EPA estimated increased cancer risks of 10^{-5}, 10^{-6}, and 10^{-7} from exposure to 300,000 fibers/liter, 30,000 fibers/liter, and 3,000 fibers/liter, respectively. Criteria and Standards Div., U.S. Envtl. Protection Agency, *Ambient Water Quality for Asbestos* C-113 (Oct. 1980). Some of this data was not available in 1975 when the court ruled, but I believe that rough estimates of about the same magnitude probably could have been made.

22. The formula is explained in William J. Nicholson, "Human Cancer Risk from Ingested Asbestos, A Problem of Uncertainty," 53 *Envtl. Health Persp.* 111, 111–113 (1983).

23. This is about the midpoint of the range found by Judge Lord in normal weather conditions, *United States v. Reserve Mining Co.*, 580 F. Supp. 11, 48 (D. Minn. 1974), and is also reasonable in light of a later epidemiological study. Eunice E. Sigurdson, "Observations of Cancer Incidence Surveillance in Duluth, Minnesota," 53 *Envtl. Health Persp.* 61, 61–62, 65 (1983).

than the level allowed by the EPA's later regulations for drinking water; the peak level was higher.[24]) With these modifications, the formula predicts about one hundred excess deaths from gastro-intestinal cancer. Alternatively, this means about 1.5 additional deaths annually in Duluth.[25] Given this risk level, the uncertain results of the epidemiological studies are not surprising: the one-in-a-thousand lifetime risk level is just on the border of being too low to detect through epidemiological studies.[26]

Was this a "significant risk"? The Supreme Court suggested in the *Benzene Case* that a one-in-a-thousand risk was significant.[27] Duluth residents were subject to more than that risk on a lifetime basis. (Also, given the disproportionately high intake of water and the rapid growth rates of children, their risk level for water-borne carcinogens may be substantially higher.[28]) In any event, one or two deaths annually in Duluth seems to be a large enough figure to justify some type of government intervention. By comparison, this is about the level of pedestrian deaths, about half the level of deaths in home fires, and about a third the number killed by drunken drivers in a typical city the size of Duluth.[29] Thus, the asbestos problem seems to be within the range of risks that we would expect public health and safety authorities to take seriously.

More recently, the EPA decided that as a general matter a life-time risk of more than one death per 10,000 heavily exposed individuals is presumptively unacceptable. The estimated asbestos risk in Duluth was ten times that level. Even taking into account other relevant criteria, such as the number of people exposed and the uncertainty of the evidence, this seems to meet

24. 40 C.F.R. § 141.62 (1998).

25. The formula is explained in Nicholson, *supra* note 22, at 111, 113.

26. John Graham et al., *In Search of Safety: Chemicals and Cancer Risk* 181 (1988).

27. *Industrial Union Dep't v. American Petroleum Inst.*, 448 U.S. 607, 655 (1979) (plurality opinion).

28. See John Wargo, *Our Children's Toxic Legacy: How Science and Law Fail to Protect Us from Pesticides* 173–199, 211, 218, (1996).

29. See Stephen Breyer, *Breaking the Vicious Circle: Toward Effective Risk Regulation* 5 (1993).

the EPA standard rather easily.[30] A summary of twenty recent EPA risk decisions does not show any examples in which a risk level this high was found acceptable.[31] So, by later regulatory standards, the estimated risk clearly qualified as significant.

The most troubling question about *Reserve Mining* is not so much the quantitative estimate of risk, but the conclusion that there was any harm at all. The evidence on this point was inconclusive at the time of the court's decision, and seemingly has remained so.[32] The EPA calculation seems designed to give the amount of risk, assuming any risk actually exists, but does not necessarily provide for the possibility that asbestos is completely harmless in drinking water. In my risk estimate earlier, I actually dealt with this problem in an indirect way. Since we don't know whether the asbestos is harmful or not, we might as well assign an equal probability (50 percent) to both possibilities. (A more sophisticated approach would be to survey experts about the distribution of probabilities, but the fifty/fifty approximation seems good enough for "home cooking.") This reasoning suggests that we ought to cut the EPA risk estimate by 50 percent to account for the possibility of complete harmlessness. I've taken this into account in my calculations by adjusting another parameter. The calculations given above in effect implicitly make this adjustment by considering only *half* of the exposed population. (I based the calculation on the one hundred thousand in Duluth, rather than the two hundred thousand living on the entire North Shore of Lake Superior.) In effect, then, I've built in a 50 percent chance that the asbestos was completely harmless. This is a crude adjustment, but should at least make the risk calculation more realistic.

Assuming that the risk counts as significant, the next question is whether land disposal was feasible. As the Supreme Court un-

30. See Janet McQuaid, "Comment: Risk Assessment of Hazardous Air Pollutants under the EPA's Final Benzene Rules and the Clean Air Act Amendments of 1990," 70 *Tex. L. Rev.* 427, 440 (1991). For further discussion of the EPA's approach, see Cross et al., "Discernible Risk," *supra* note 2, at 61, 65–73, 80–81.
31. McQuaid, *supra* note 30, at 460–461.
32. See 50 Fed. Reg. 46961 (1985).

derstands it, this means a determination of whether it was economically and technologically possible to accomplish land disposal.[33] The Eighth Circuit clearly thought so. In retrospect,
the judges were obviously right, since the company managed to
make the change. (Though the company did later collapse fi-
nancially, there doesn't seem to be any evidence of a causal connection.) Thus, the *Reserve Mining* decision seems to be correct,
given the Supreme Court's later interpretation of significant risk
and feasibility.

On its surface, this kind of feasibility analysis looks very different from a cost-benefit analysis. But the difference may not
be quite as complete as it appears. In reality, a decision maker is
very likely to take a peek at costs when deciding whether a risk
is significant. What is "feasible" to control a major risk might be
considered infeasible when the risk is much smaller.[34] The result
may be that costs and benefits are really being compared, though
only covertly. Without this covert balancing, the results of the
feasibility test could be bizarre. Is a risk of one in a million significant? Probably not, but what if eliminating the risk will cost
only a few dollars? On the other hand, the EPA generally considers a risk of one in ten thousand to be significant, but if the cost
is measured in billions, we might want to reconsider just how
significant we consider the risk. In theory, the EPA is supposed
to consider cost and risk quite independently, but it would be
irresponsible to ignore the linkage. Whatever consideration of
the relationship between costs and benefits that takes place,
however, cannot be an explicit part of the feasibility approach.

The concern here relates more to the process of decision than
to the final outcome. In the end, the feasibility approach obviously gives greater weight to environmental benefits than to
costs, which may well be appropriate.[35] The problem is not the

33. 452 U.S. at 509.
34. Similarly, in ordinary life, what is feasible depends on the context. It
would be feasible for someone to rush to the airport and travel across the country
for a family emergency. But we would not say that doing the same thing was a
"feasible" method of returning a library book at the last minute.
35. In this sense, feasibility analysis is like "compelling interest" analysis in
constitutional law. (For example, the government can regulate certain kinds of

ultimate balance, but the fact that the balancing is forced under-
ground, rather than being explicit. In a democracy, policy decisions should not be concealed from the public.

Another problem with feasibility analysis is that it varies the
degree of regulation with the economic health of the industry.
Hence, feasibility analysis gives fear of bankruptcies and unemployment (as opposed to the total dollar value of compliance
costs) more weight than an economist might prefer. In principle,
it is unclear why an industry should be exempted from compliance simply because it is in financial difficulty. Just because its
technology is inefficient, or its other costs are rising, or demand
for its product is falling, why should an industry be allowed to
impose a health risk on the public? It seems odd to punish successful industries by imposing harsher regulations, while giving
unsuccessful industries exemptions. Whatever might be said on
this subject in theory, however, the practical political answer is
clear. Regulatory agencies dread the specter of plant closings,
which are every politician's nightmare, and therefore every bu-
reaucrat's.

Cost-Benefit Analysis and Its Discontents

The main disadvantages of feasibility analysis are that it may lead
us to favor one side of the balance excessively and that its rhetoric may conceal the trade-offs that we are really making. Its major competitor, cost-benefit analysis, is an effort to make trade-
offs explicitly and even-handedly.

In place of the intuitive balancing that sometimes takes place
in the guise of the "feasibility" test, cost-benefit analysis purports to offer a precise, scientific way of comparing costs and
benefits. As we saw in chapter 2, the key to this comparison is
to place a dollar value on the benefits of regulation. Thus, in
applying cost-benefit analysis to public health problems like *Re-*

---

harmful speech, but it must show that the regulation is necessary to achieve a
compelling government interest, not merely a favorable cost-benefit ratio.) This
test still allows some degree of balancing, but in a structure that gives a decided
preference to one side of the balance. See Daniel Farber, *The First Amendment*
32–33 (1998).

serve Mining, economists must include some measure of the value to be attached to human life.[36]

Many people find the idea of putting monetary values on lives to be intrinsically offensive.[37] Rather than speaking of the cash value of a life, we might better speak of the amount that we collectively and individually are willing to sacrifice in order to save lives. We may want to pretend that this amount is infinite, but the harsh reality is that there are limits to the resources we can or should devote to safety.[38] We don't require people to wear safety helmets to watch TV or to use seat belts on park benches, even though someone might occasionally be saved from a bad fall by these precautions. Complete safety is a chimera. At some point, we are simply unwilling to pay the price for greater safety.

As mentioned in the previous chapter, economists usually look to the amount people are willing to sacrifice to reduce the amount of risk in their own lives. The best information comes from labor markets. People have to be paid extra to take dangerous jobs, and this provides a measure of how much they value their lives.

This data is useful, but it would be a mistake to rely on it unquestioningly. We have reason to question the risk preferences revealed by industry wage patterns.[39] Studies by cognitive psychologists show that individuals are quite bad at estimating risk levels.[40] Even where a risk is known, people do not process the

36. Clayton Gillette and Thomas Hopkins, *Federal Agency Valuations of Human Life: A Report to the Administrative Conference of the United States* 1 (1988).

37. See id. at 24–26 (analyzing the ethical difficulties in putting a monetary value on human life).

38. See id. at 1–4 (discussing different life values assumed by federal regulations designed to save lives).

39. Id. at 41–49. As Gillette and Hopkins point out, a number of biases created by the realities of labor markets prejudice the accuracy of this approach. These include the absence of meaningful employment options, the lack of information available to workers, and the presence of externalities whereby some of the costs associated with a given risk are not borne by the worker or the worker's immediate family. See also Gillette and Krier, *supra* note 7, at 1038–1042.

40. See Talbot Page, "A Generic View of Toxic Chemicals and Similar Risks," 7 *Ecology L.Q.* 207, 225–228 (1978). While part of the problem may be attributed to limited knowledge, research indicates that people systematically assign insufficiently low probabilities to rare events. Id. at 226–227 (citing experiments by Alpert and Raiffa and by Tversky and Kahneman).

information very well. They miscalculate in considering combinations of risks, they ignore background information when they are assessing new data, and they are easily swayed by trivial changes in the presentation of information.[41] For example, they will favor a medical option when told it has a 20 percent chance of saving their lives, but shun it when told it has an 80 percent chance of failure, though the two are equivalent. Finally, when using labor information as a basis for assessing risk preferences, as is commonly done, we need to be especially cautious. The operation of labor markets is still poorly understood by economists, which makes interpretation of the data difficult.[42]

The upshot is that the methods economists use to determine risk preferences are quite imperfect. Not surprisingly, the results are extremely variable, with estimates of the value of life ranging from $15,000 to $3 million. Substantially higher values can reasonably be justified for involuntary risks.[43] Kip Viscusi, an economist at the Harvard Law School who is a leading advocate of

---

41. This research is summarized in Amos Tversky and Daniel Kahneman, "Rational Choice and the Framing of Decisions," in *Rational Choice: The Contrast between Economics and Psychology* 67 (Robin Hogarth and Melvin Reder eds., 1987).

42. See Thomas Kniesner and Arthur Goldsmith, "A Survey of Alternative Models of the Aggregate U.S. Labor Market," 25 *Journal of Economic Literature* 1241, 1272–1273, 1275 (1987) (discussing the uncertain state of the knowledge concerning the relationship between wages and labor supply). See also Robert Hutchens, "Seniority, Wages and Productivity: A Turbulent Decade," *J. Econ. Persp.*, Fall 1989, at 49, 49, 53–54 (1989) (stating that the relationship between seniority and wages is presently a puzzle to economists).

43. See Thomas McGarity, *Reinventing Rationality: The Role of Regulatory Analysis in the Federal Bureaucracy* 146 (1991) (agencies use figures from $340,000 to $7.5 million); E. J. Mishan, "Consistency in the Valuation of Life: A Wild Goose Chase?" in *Ethics and Economics* 152, 160–161 (Ellen Frankel Paul et al. eds., 1985). The discrepancy in the cost effectiveness of safety regulations is even greater, ranging from $132 million dollars per life saved for the 1979 regulation of diethylstilbestrol (DES) in cattle feed to $200,000 per life saved for protection from airplane cabin fires. Richard Zeckhauser and W. Kip Viscusi, "Risk within Reason," 248 *Science* 559, 562 (1990). To a large degree, such discrepancies can be attributed to different risk-cost assumptions made by the various federal regulatory agencies. For example, Federal Aviation Administration regulatory policy considers the cost effectiveness of the regulation on the basis of earnings lost through accidental deaths, while (theoretically at least) the EPA and the FDA regulate without consideration of cost. Id. at 562–563.

this methodology, concludes after a careful consideration of the possible distortions in the empirical studies that the most reasonable recent estimates are clustered in the $3 to $7 million range.[44] Interestingly, other studies show that unions typically bargain for safety costs of around $5–6 million per life saved.[45] This provides some reassurance against the fear that the occupational studies merely reflect the lack of bargaining power or information available to individual workers.[46] Apparently, even when workers have substantial bargaining leverage, they make similar trade-offs.

In using these figures, we should keep in mind the caveats in the current Office of Management and Budget (OMB) guidelines on cost-benefit analysis. According to OMB, "[U]se of occupational-risk premiums can be a source of bias because the risks, when recognized, may be voluntarily rather than involuntarily assumed, and the sample of individuals upon which premium estimates are based may be skewed toward more risk-tolerant people." Consequently, administrative agencies are instructed that using the figures in environmental cases "may not be entirely appropriate," and they are told to provide an explanation for their selection of estimates and adjustments to those estimates.[47] As economists from Resources for the Future have explained, "There is a growing recognition that the compensating wage studies have limitations for valuing death-risk reductions in an environmental context." These economists point out several additional limitations of such studies for valuing environmental risks: "they reflect risk preferences of perhaps a less-risk averse group than the average in society; they reflect voluntarily born risks; more life-years are lost to accidental death than to, say, cancer . . . ; and the source of the risk is an accident, rather than,

44. W. Kip Viscusi, *Fatal Tradeoffs: Public and Private Responsibilities for Risk* 73 (1992).

45. Breyer, *supra* note 29, at 22.

46. This fear is expressed in Elizabeth Anderson, *Value in Ethics and Economics* 196 (1993).

47. Regulatory Working Group, *Economic Analysis of Federal Regulations Under Executive Order 12866* (Jan. 11, 1996) <http://www.whitehouse.gov/WH/EOP/OMB/html/miscdoc/riaguide.html>.

say, a business that pollutes as part of its normal operations."[48] To muddy the waters further, there is evidence that people place a significantly higher value on avoiding death from cancer than from other causes.[49]

In short, the "value-of-life" numbers should probably come with a warning sticker: "SUBJECT TO GREAT UNCERTAINTY. USE WITH CARE. PROFESSIONAL ASSISTANCE STRONGLY ADVISED." Still, these figures are probably the best benchmark we have about what value people put on avoiding risks in everyday life.

The value-of-life figure illustrates the role intangible value judgments play in cost-benefit analysis. I find the $1–10 million figures plausible, but not because I consider myself technically competent to assess the methodologies of the various studies. In part, I find the figures plausible because they are supported by able economists like Viscusi, who seem unlikely to have any bias in favor of inflating the results. But I also find the figures plausible because they are a reasonably good fit with my own intuitions about how society should value human life. There used to be a popular TV show called *The Six Million Dollar Man,* about a bionic person. The implication of the title was that a person who was worth $6 million was pretty valuable, and I have the same sense about these value-of-life figures. A figure in the $1–10 million range seems high enough to take the value of life seriously. It is not, for example, out of the range of jury awards in wrongful death cases, and it is an amount that we might seriously consider funding to save a life through some expensive medical procedure. On the other hand, a figure that was much higher would seem a bit extravagant. For example, a figure of $50 million would imply that people would be willing to spend $50,000 apiece to escape a one-in-a-thousand risk of death, which seems implausible. So the $1–10 million figure strikes me as appropriate for reasons having little to do with econometric studies, and more to do with personal value judgments.

48. Raymond Kopp et al., *Discussion Paper: Cost-Benefit Analysis and Regulatory Reform: An Assessment of the Science and the Art* 20–21 (Resources for the Future, Jan. 1997).

49. See Cass Sunstein, *Free Markets and Social Justice* 310 (1996) (contingent valuations about 50 percent higher for cancer as opposed to accidental deaths).

Cost-benefit analysis provides an aura of certainty. That must be counted as in some sense an advantage—it is painful to make difficult decisions about vital issues, and it is therefore reassuring to have the results underwritten by scientific certainty. But as economists themselves are well aware, the aura of certainty is illusory. Cost-benefit analyses involve many judgment calls, which do not render the analyses worthless but do introduce discretion.

In performing a cost-benefit analysis in a situation like *Reserve Mining,* at least three discretionary judgments must be made. The first point of discretion is the choice of a value of life, as we've just seen. In *Reserve Mining* the range of defensible choices runs from around $1 million up to about $10 million, since this is an involuntary risk.

The second point of discretion is appraising the risk. How conservative (risk adverse) should we be in estimating risk levels? The best argument for conservatism is that mortality rates alone do not fully define risk. In *Reserve Mining,* other factors are important: the risk was novel, it involved cancer, and a whole community was at risk. Moreover, the mortality estimates themselves were highly uncertain.[50] Since the risk assessment was based on occupational exposures, it may underestimate the risk to children, who are more susceptible to carcinogens. There is also a sort of primal fear of tampering with water supplies (even today the phrase "poisoning the well" carries a clear message). In short, it was a somewhat scary situation. The decision makers may want to hedge against the possibility that the best current estimate of risk will turn out to be too low. Thus, they may well take a broader view of risk than that taken by actuaries.

The third point of discretion is the discount rate, if any, for future deaths. Assuming about a hundred deaths, land-based disposal is justified, without discounting, at any value of life over $2 million. Considering that the risk imposed is involuntary, a $2 million value-of-life figure seems quite reasonable. Thus, if we don't discount, the court was clearly correct in ordering land

---

50. For a discussion of these factors, see Richard Pildes and Cass Sunstein, "Reinventing the Regulatory State," 62 *U. Chi. L. Rev.* 1, 55–64 (1995).

disposal. Discounting could change this conclusion. Relatively small changes in interest rates can make big differences in final results. A dollar spent a year from now is less significant than a dollar spent today because the year's delay provides the opportunity to earn interest in the meantime. On similar grounds, many economists believe that a life saved a year from now should weigh somewhat less than a life saved today.[51] In *Reserve Mining*, for example, future cancer deaths would be spread over a long period, and their weight would be diminished accordingly. Economists disagree, however, about what interest rate should be used to calculate the present value of future deaths. We will explore this topic in depth in chapter 5. Suffice it to say that there is considerable ground for dispute about the appropriate discount rate. The result of the cost-benefit analysis could turn on this question.

Using the EPA's 1983 risk estimate, I've done some "back of the envelope" cost-benefit calculations for *Reserve Mining*. Obviously, I'm not pretending to give accurate final numbers; the purpose of this tentative effort is only to show the range of possibilities. My rough calculations show the sensitivity of the results to these three parameters (value of life, risk level, and discount rate). It turns out that the calculation can come out either way, depending on how the three discretionary decisions are made. Essentially, if a conservative choice is made for any two of the three, the cost-benefit analysis comes out in favor of land disposal. For example, land disposal is apparently justified if we assume a 5 percent discount rate for future deaths, the EPA risk estimate, and a value of life of $8 million.[52] If we assume a lower discount rate or make a higher estimate of risk, we can reach the same result with a lower value-of-life figure. Thus, land disposal

51. See Gillette and Hopkins, *supra* note 36, at 56–57.
52. The company indicated that the estimated $240 million would be spent on land disposal. We are assuming 1.5 lives saved per year, at a value of $8 million each, or $12 million per year. The discounted value of an infinite stream of $1 per year is $20 because $20 produces $1 of interest annually. This gives a total present value of $240 million. This exceeds the cost of regulation, taking into account that that cost will accrue over several years and must also be reduced to present value.

is also justified with the EPA risk estimate and a $2 million value of life if the discount rate for future deaths is 1 percent.[53] Alternatively, we can decide to be more conservative in our risk estimate than the EPA. If so, the other two figures can be lower.

What I have presented is an extremely crude cost-benefit analysis. The treatment of compliance costs is quite unsatisfactory. It takes the company at its word about the extent of the costs, which is often likely to lead to overestimation. It also ignores the possible health and environmental impacts of land disposal. Depending on how the discounting issue is analyzed, it may also be important to consider how the compliance costs are being financed and whether the effect is to eliminate other investments that might have increased future productivity. Finally, other effects of the dumping are simply ignored: no effort is made to account for possible ecological damage, for health risks due to air pollution by asbestos, for potential recreational uses of the area, or for the nonuse values that might be measured through a contingent valuation. But the outcome of the cost-benefit analysis seems close enough to show that the ultimate result of a more complete analysis probably could go either way, depending on the specific choices made by the analyst.

What this means is that, in *Reserve Mining* at least, the methodology of cost-benefit analysis probably does not in itself determine the result. The result is actually driven by the discretionary policy decisions made in implementing the analysis. If we make highly risk-averse implementation decisions, regulation is justified; if we resolve those decisions in a more risk-accepting manner, the decision goes the other way.

Most of the debate about cost-benefit analysis seems to be based on an assumption that the technical analysis does provide firm answers—the dispute is only about whether they are the answers to the right questions. But in practice, perhaps cost-benefit analysis doesn't provide final answers at all. What it may provide often is a framework in which policy decisions about

53. The present value of annual payments of $1 per year is $100 at a 1 percent discount rate. See Richard Brealy and Stewart Myers, *Principles of Corporate Finance* 30–33 (1981). So we have a total value of lives saved of $1.5 \times 2 \times 10^6 \times 100$, or $300 million.

public risk-aversion can be made and applied. Of course, use of this framework may have the effect of molding the decision maker's approach to a problem, but it may not necessarily foreclose particular outcomes.

Just as cost-benefit analysis has built-in flexibility, so does feasibility analysis. Consider *Corrosion Proof Fittings,* probably the clearest judicial embrace of cost-benefit analysis. The court's endorsement of cost-benefit analysis may not have been key to the outcome. Even under the feasibility approach, the decision might well have been the same anyway. Recall that the decision overturned a nationwide ban by the EPA on asbestos products. The court rejected parts of the asbestos ban as addressing only trivial risks, no more socially significant than the number of deaths caused by swallowing toothpicks. If it had been applying a feasibility analysis, the court clearly would have continued to find these asbestos risks insignificant. Other asbestos risks were clearly significant, but the remedy raised serious questions in the court's mind because of doubts concerning the availability of safe alternatives to asbestos for these uses. Concentrating on these subindustries, the court could have phrased its conclusion in terms of feasibility analysis: the agency had failed to carry the burden of showing that elimination of asbestos was feasible with respect to these subindustries.

Given the court's evident skepticism toward the regulation, and lack of respect for the agency, it is hard to believe that switching from cost-benefit analysis to feasibility analysis would have had a major impact on the result. Sometimes attitude counts for more than technique.

Much of the discussion about environmental risks has been dominated by the dispute between advocates of cost-benefit analysis and feasibility analysis. In large part, the assumption of the debaters seems to be that the distinction is outcome determinative. While the choice between these methodologies may well help to shape outcomes, it would be a mistake to think that either methodology is capable of producing a hard, definitive answer, even within its own framework. Both methodologies are handicapped by uncertainties about the levels of risk. Although we are in a somewhat better position to identify and even quantify risks today than at the time of *Reserve Mining,* major uncertainties remain, as I discuss in chapter 6. Both forms of analysis are also flexible enough to allow considerable discretion. Feasibility analysis is flexible because critical terms such as "significant" risk and "feasible" controls are hard to define. Cost-benefit analysis is flexible because uncertainty surrounds crucial parameters such as the value of life and the discount rate. Whatever approach we do adopt, judgment calls and value judgments will be required.

The conflict between cost-benefit analysis and feasibility analysis is more subtle than one might expect. Neither can be considered illegitimate in principle. Neither guarantees a firm answer, right or wrong. Both provide frameworks, and the choice of the right framework must be a pragmatic one. We must ask what framework works best—best in terms of the limits on available information, best in terms of political fairness and accountability, and, most important, best is terms of capturing our society's fullest understanding of the values at stake. Selection of a particular method of analysis may not force us to a particular conclusion, but it may at least give us a strong nudge in one direction.

Our quest should not be for a foolproof method to decide environmental issues. Rather, we need a mode of analysis that allows us to incorporate our values as intelligently as possible. Lack of a computer program for decision making is, of course, no cause for anxiety among pragmatists, who have little expectation that hard decisions can be made easy by the correct choice of methodology. But having said this, the major question remains: How should eco-pragmatists tackle difficult problems such as the asbestos threat in *Reserve Mining?* In the next chapter, we begin to tease out some answers, focusing first on the appropriate roles for feasibility analysis and cost-benefit analysis.

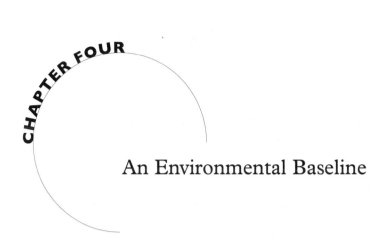

CHAPTER FOUR

An Environmental Baseline

Given my belief in legal pragmatism, it should be no surprise that I see little hope for a purely mechanical method of deciding these difficult issues. There is no escape from making hard judgments, both about technical matters and about social values. But my vision of pragmatism recognizes the usefulness of overarching principles, which can help guide judgment in particular cases. This aspect of pragmatism, however, is at odds with popular notions. To many people, pragmatism is interchangeable with ad hoc balancing: a "pragmatist" is someone who weighs all the factors (probably with an emphasis on materialistic costs and benefits) and then tries to make a good decision in any given case. This might be most easily accomplished through a form of "soft" cost-benefit analysis, which would compare costs and benefits without attempting to quantify every factor. In the environmental area, the most thoughtful proponent of such a balancing approach is Cass Sunstein. Analysis of his views will help clarify the correct understanding of pragmatism in environmental law.

Combining cost-benefit analysis with an intuitive weighing of less tangible values is admittedly appealing. It promises to combine economic analysis with other intangible values that "number crunching" may slight. It bespeaks an open-minded willingness to consider each case on its own merits, without biases or preconceptions. Nevertheless, I will argue, this is the wrong approach to environmental policy. Our society has basic commitments, including one to environmental quality, and those commitments should form the baseline for analysis. These commitments are not boundless, and cost-benefit analysis may identify when continued attachment to them would be unreasonable in a particular case. But rather than approaching each case anew, we should start from the environmental norms that our society has unmistakably embraced.

In concrete terms, this analysis translates into strong reliance on the feasibility approach to regulation. This approach incorporates our societal commitment to environmental protection by requiring the strongest feasible efforts to obtain environmental quality. Besides being limited by notions of feasibility, environmental protection may also need to be qualified in another way. Although "feasible" in some sense of the word, achievement of an environmental goal may sometimes involve costs that are grossly disproportionate to any plausible benefit. Thus, cost-benefit analysis may serve as a useful backstop for feasibility analysis to handle these situations. We should always begin, however, with a presumption in favor of protecting the environment except where infeasible or grossly disproportionate to benefits.

A "Kinder and Gentler" Cost-Benefit Analysis?

The call for cost-benefit analysis begins with the recognition of the inevitability of tradeoffs between cost and environmental quality. Cost-benefit analysis promises a mechanical technique for making these tradeoffs, relieving us of the burden of difficult value judgments. But sophisticated advocates realize that cost-benefit analysis cannot deliver on this promise. There are too many judgment calls and too many unquantified factors inherent in environmental problems. Hence, unquantified factors must also play some role, though advocates of cost-benefit analysis, like the court in *Corrosion Proof Fittings*, hope that this role will be a secondary one.

Sunstein's work reveals some of the tensions inherent in this approach. He has been a long-time critic of environmental regulation as clumsily designed and poorly implemented, and among the remedies he has long supported is cost-benefit analysis. Yet his more jurisprudential work, as we saw in chapter 2, opposes the kind of economic reductionism underlying formal cost-benefit analysis. Not surprisingly, he has recently retreated from staunch support for formal cost-benefit analysis toward a more nuanced view.

Sunstein has recently called for a modified form of cost-benefit analysis, which he has formulated in different ways. As we saw in chapter 2, he believes that public values rather than private preferences should drive government decisions. Given his philosophical critique of key concepts of welfare economics like the "willingness to pay" standard, he obviously does not endorse economic efficiency as the sole basis of regulatory policy. But he does call for expanded use of cost-benefit analysis at the expense of more protective approaches to environmental regulation. In one formulation, officials would engage in a two-stage decision process. The first stage would consist of a quantitative cost-benefit analysis; the second would introduce other values, "if any are relevant," that cost-benefit analysis leaves out.[1]

Another formulation distinguishes between different types of statutes. According to Sunstein, some statutes such as federal pesticide and toxics regulations are designed to overcome "market failures" such as "an absence of sufficient information on the part of consumers, harms to third parties, or collective action problems of various sorts." For these statutes, Sunstein says, "there is much to be said" in favor of a quantified cost-benefit analysis, though even for these statutes, "goals other than those rooted in economic efficiency may legitimately bear on the decision." Other statutes, which currently are considered "absolutist," could be amended to allow balancing.[2] But Congress should "understand costs and benefits in economic terms only for statutes that are designed to overcome market failures." This economic analysis could be qualified when "the risk at issue is inequitably distributed, and when political actors believe it deserves special attention for that reason."[3]

---

1. Cass Sunstein, *Free Markets and Social Justice* 139 (1996).
2. Id. at 369–370.
3. Id. at 369. For an environmentalist critique of Sunstein's proposal, see Thomas McGarity, "A Cost-Benefit State," 50 *Admin. L. Rev.* 1 (1998).

When a statute is not based on market failure, Sunstein says, "Congress should still require cost-benefit balancing as a general background rule; but it should understand the definition of costs and benefits to be sufficiently wide open as to allow administrators to depart from purely economic criteria." Hence, Sunstein calls for a cluster of regulatory reforms: a general requirement of cost-benefit balancing, with costs and benefits understood in an open-ended sense; a mandate of cost-effectiveness for all regulations; and quantitative cost-benefit analysis where "there is a considered legislative judgment that the statute is a response to a market failure, economically defined." In connection with these reforms, Congress would set benchmark values of $3–10 million per life saved.[4]

Sunstein's view deserves careful consideration not only because of his stature as a legal scholar, but also because it is the most nuanced version of cost-benefit analysis. What Sunstein proposes is a moderate, humane form of cost-benefit analysis, a distinct improvement over the blunter versions proposed in Congress. Yet even this version is flawed. It can be criticized for leaving agencies too much discretion. It also can be criticized for failing to appreciate the extent to which even the most rigorous cost-benefit analysis involves recourse to nontechnical value judgments. It is more difficult than it may seem to split a decision into a purely technical quantitative phase and a more open-ended qualitative phase. The most fundamental difficulty, however, is the proposal's failure to acknowledge the nature of our national commitment to the environment.

By approaching problems in terms of balancing, whether in the form of cost-benefit analysis or a more intuitive comparison of competing values, Sunstein suggests a stance of presumptive neutrality on the part of the decision maker. In one formulation, the decision maker actually begins by using the "neutral" technique of cost-benefit analysis and only then considers the impact of other values. For statutes addressed to market failures—a category that Sunstein does not define, but that seems to include toxic regulation and pollution laws, judging from his examples—Sunstein seems to think that "other" values will enter the analy-

4. Id. at 371. Richard Pildes collaborated with Sunstein on this analysis.

sis only in exceptional cases. For other statutes, they will play a greater role. But in any of Sunstein's formulations, the decision maker begins by open-mindedly assessing the values on each side of the balance, rather than beginning with a bias in either direction.

Lack of bias and openmindedness do seem like undeniable virtues, about which it is hard to complain. But the whole point of having commitments is that one does not enter every decision with an open-minded willingness to examine all competing considerations on an equal footing. Instead, one begins with a different baseline—a baseline in which certain values are presumptively controlling and can be upset only under unusual circumstances. The commitments now embedded in federal law generally take an environmentalist baseline, with a presumption in favor of environmental protection. Sunstein would abandon this baseline for a more detached stance. The result would be to designate economic efficiency as the presumptive outcome, shifting the burden to those who advocate other values.[5]

Even by the time of *Reserve Mining*, over twenty years ago, it was clear that we had moved beyond case-by-case balancing in environmental decision making. By then, Congress had passed the Clean Water Act, the Clean Air Act, and the Wilderness Act. All of these statutes set rigorous environmental goals and make only limited (though significant) provision for considering cost. In retrospect, Congress might have been unrealistic in failing to provide for broader consideration of economic cost in setting the basic goals. But Congress did make it unmistakable that environmental quality is a pressing national goal. Case-by-case balancing had failed, and a firmer sense of direction was called for. Environmental regulation has now become more sophisticated and further adjustments may well be necessary, but the basic

5. In the end, of course, burdens of proof and presumptions are not always decisive. In many situations, Sunstein's proposal might lead to the same results as the environmentalist alternative. But the difference in baselines is nonetheless important. When the balance seems close or insufficient information is available, the initial baseline is likely to control. Moreover, how we resolve difficult environmental problems may partly depend on the mindset with which we approach them. For these reasons, it is important to have a baseline that does justice to our society's strong commitment to environmental values.

commitment to environmental quality has stood the test of time. Sunstein's proposal would be a step backwards toward relinquishing this national commitment. As I will argue below, there is nothing inevitably superior about the kind of "neutral" approach he advocates. However we approach environmental problems, our society cannot avoid the necessity of making value judgments. A decision to count the interests of polluters and their victims equally is simply another value judgment—and to my mind, an unappealing one. Of course, this is no argument for discounting the polluter's interests altogether; it is only an argument for giving those interests a subordinate position.

To make the case for a pro-environmental baseline, I will begin by discussing the objections to the neutral baseline. This baseline would have us start over from scratch in making each environmental decision, balancing the interests as if we had never before encountered an environmental issue. Given his belief in collective deliberation, it is ironic that Sunstein's baseline fails to keep faith with our sense of ourselves as a community. After critiquing Sunstein's views, I will then turn to the task of sketching the pro-environmental baseline and considering the role that cost-benefit analysis can properly play within that baseline. My suggested approach would require that we eliminate significant environmental risks unless the benefits are clearly out of line with the costs. Succeeding chapters will consider some other aspects of the environmentalist baseline—in particular, how to respond to the long-range and highly uncertain nature of environmental problems.

The Case against Neutrality

One of Sunstein's jurisprudential insights has been the importance of baselines in legal thinking. As he recently pointed out, there is no way short of anarchy to avoid making an initial assignment of entitlements, which inevitably shapes later legal developments.[6] In his analysis of environmental law, however, he has failed to make full use of this insight.

In making the argument against a neutral baseline, I will show

---

6. Sunstein, *supra* note 1, at 17.

how baseline determinations are central to understanding any approach to environmental regulation, whether it be a conventional cost-benefit analysis, Sunstein's more open-ended balancing, or the feasibility approach of so many environmental statutes. An environmentalist baseline is warranted, I contend, notwithstanding Sunstein's contrary arguments and the claims made for neutrality by Ronald Dworkin and others. I will use a leading case from the law of nuisance to illustrate why a neutral baseline is unacceptable.

Choosing Baselines

It may be helpful to begin by seeing how even cost-benefit analysis implicitly assumes a baseline. For convenience, in previous chapters I have talked about "willingness to pay" as the basis for cost-benefit analysis. As a practical matter, this is normally the standard used in cost-benefit analysis. In theory, however, another standard is available: "willingness to accept." Under this standard, instead of asking what people would pay to get cleaner air, you ask what price they would demand before agreeing to accept air pollution.

An example might help clarify the distinction. Consider the decision to destroy a pod of whales. If whalers have the legal right to kill the whales, environmentalists might not be willing (or able) to pay the companies enough to get them to stop. A cost-benefit analyst would say that company profits were greater than the harm to the environmentalists, so whaling would be economically efficient. Thus, the whalers can claim that their actions meet the "market" test of cost-benefit analysis: in a world of perfect markets, the whaling would proceed because its benefits to them outweigh the costs to the environmentalists.

The whalers' argument covertly assumes, however, that they own the entitlement to control whaling. If the environmentalists had the legal right to prevent whaling, they might demand a much higher price to sell that right to the whaling companies. (One reason for the disparity is that environmentalists are in a sense "wealthier." They would have a legal entitlement they didn't own before, and changes in wealth shift the demand curve.) So if the environmentalists own the entitlement, the cost-benefit analysis may well show that whaling is inefficient. We can't decide whether the whaling is economically efficient until we know who has the entitlement. Thus, cost-benefit analysis is indeterminate in this situation, and we have to look elsewhere to decide whether we should allow the pod of whales to be killed.

This is not merely a hypothetical issue. Contingent valuation studies have shown that there is actually a large difference in many situations between willingness to pay and willingness to accept. Perhaps "buy low/sell high" is a deeply engrained impulse. In any event, the willingness to accept (WTA) figure is often twice as high as that for willingness to pay (WTP). The reasons are unclear. The disparity could be due to wealth effects and other economically rational considerations. Alternatively, noneconomic factors may be the cause, not unlike the way people view an 80 percent chance of death differently than the equivalent 20 percent chance of survival. Perhaps people feel irrationally attached to the status quo and are reluctant to surrender current entitlements (the so-called endowment effect).[7] Perhaps they feel that selling away their rights would make them accomplices to environmental destruction, while simply failing to pay for preservation would make them feel less guilty. Or perhaps, after all, they really are exhibiting a subtle form of economic rationality. Some analysts argue that the difference in results can be explained on purely economic grounds, particularly for unique public goods with no close substitutes and with high demand.[8] Whatever the reason, the difference between WTA and WTP is substantial.

As a practical matter, economists performing contingent valuations strongly prefer not to use WTA because they often get very high or infinite prices or outright refusals to sell. This problem leads to a dilemma: "[A]sking people to accept payment for a degradation in the quantity or quality of a public good simply does not work in a CV [contingent valuation] survey under many

7. Id. at 51–53, 248–253.

8. See Robert Mitchell and Richard Carson, *Using Surveys to Value Public Goods: The Contingent Valuation Method* 30–38 (1989); Daniel Levy and David Friedman, "The Revenge of the Redwoods: Reconsidering Property Rights and the Economic Allocation of Natural Resources," 61 *U. Chi. L. Rev.* 493, 507–517 (1994).

conditions, yet substituting a WTP format where theory specifies a WTA format may grossly bias the findings."[9] Thus, when using WTP, cost-benefit analysts are implicitly making a particular assignment of baselines. This is a choice that often tilts the table against environmental protection.[10]

Cost-benefit analysis purportedly gives equal weight to the interests of both sides and is therefore unreceptive to the use of moral rights as an analytic tool. For example, Ronald Coase objects to the notion that someone making noise should be seen as invading the rights of a neighbor. Coase, a University of Chicago economist who won the Nobel prize, argues that the neighbor can equally well be seen to be invading the noisemaker's rights with his demand for quiet.[11] For Coase, the only relevant question is which party would be willing to pay more for his preferred outcome.[12] In Coase's analysis, regulation can be conceptualized as mimicking an auction of unallocated entitlements to the highest bidder.

To see how this neutral baseline operates, consider the *Reserve Mining* situation. A cost-benefit analysis essentially estimates the harm done by the asbestos pollution and considers what the Duluth victims would be willing to pay to eliminate that harm. Against this, we balance the cost of switching to land disposal. We could accomplish much the same result by organizing an auction in which the city of Duluth would bid against the company for "ownership" of this portion of Lake Superior's water. The auction is completely fair in the sense that each side has an equal opportunity to prevail and the one with the greatest stake will win.

But an auction is not the only possible approach. Alternatively,

9. Mitchell and Carson, *supra* note 8, at 32–33, 37.

10. See Jack Knetsch, "Environmental Policy Implications of Disparities between Willingness to Pay and Compensation Demanded Measures of Values," 18 *J. Envtl. Econ. & Mgmt.* 227, 230–235 (1990).

11. Ronald Coase, "The Problem of Social Cost," 3 *J.L. & Econ.* 1, 8–10 (1960).

12. In the absence of transaction costs, Coase believes that the parties will bargain their way to this result anyway. Only where transaction costs impede bargaining will the legal rule be decisive. Sunstein believes that the Coase Theorem is "in some cases" erroneous. Cass Sunstein, *After the Rights Revolution: Reconceiving the Regulatory State* 250 n.38 (1990).

we could give the entitlement to the company and require the city to use its power of eminent domain. Or we could give the entitlement to Duluth and require the company to buy out the city (the WTA standard). Either of the latter alternatives gives one side or the other an "edge," whereas the auction treats them equally—and therefore, one might conclude, justly.

It may seem reasonable enough to treat entitlements as unallocated and to use an auction to allocate them. Nevertheless, in some contexts, both steps strike many people as highly unnatural, if not immoral. Viewing entitlements as unallocated seems reasonable enough in the abstract, but not when the question is whether the entitlement to control sexual intercourse belongs to a woman or her would-be rapist. Even where entitlements really *are* unallocated, as in the case of an orphan who must be placed for adoption, an auction to award the child to a particular couple is hardly the most readily accepted mechanism. It is not unheard of to attempt an economic analysis of issues such as rape and adoption. The best known efforts are those of Richard Posner, a founder of the "law and economics" movement who is now a distinguished federal judge. Judge Posner's use of a "neutral" baseline made his analyses of both these situations controversial, if not shocking to many people's moral sensibilities.[13]

Like Posner, Sunstein also takes a neutral stance, though a softer one. He begins by taking bids from the contesting interests in order to determine willingness to pay. For statutes based on market failures—a category that he leaves vague, but that seems to include most pollution statutes—the bidding normally determines the regulatory outcome. Even in these situations, and to a greater extent under other statutes, the decision maker then opens the floor to arguments based on other public values. In a few situations, perhaps involving endangered species or other special public values, the bidding stage may be skipped altogether.

The process is much more open-ended, especially for nonpol-

---

13. See Richard Posner, "An Economic Theory of the Criminal Law," 85 *Colum. L. Rev.* 1193, 1198–1199 (1985) (economic analysis of rape); Richard Posner, "Adoption and Market Theory: The Regulation of the Market in Adoption," 67 *B. U. L. Rev.* 59 (1987).

lution statutes, than Posner's efficiency test. But the decision begins on an even footing. No one is entitled to say: "I have a right to a safe environment, and the burden is on anyone who seeks to infringe that right." Nor are the regulated parties entitled to say: "I have the right to use my property as I see fit, and the burden is on anyone who seeks to modify that property right." There is a level playing field.

For any regulatory scheme, we have a choice of at least three baselines. First, we can use the common law as a baseline, which usually means that the regulated party holds the presumptive entitlement. (An alternative that often leads to the same result is to take the status quo as the baseline, since the preregulatory status quo is often the common law.) Sunstein rightly says that the common law baseline is often inappropriate in the post–New Deal world. Second, we can use a "neutral" normative baseline, which treats entitlements as initially unassigned, so that neither party begins with any presumptive claim. Third, we can assign the presumptive entitlement to the beneficiaries of the regulatory program. Congress has treated environmental risks as impermissible except when required by considerations of feasibility. Rather than cost-benefit analysis, Congress has adopted a pro-environmental baseline for the control of air and water pollution,[14] carcinogens in the workplace, and hazardous waste sites,[15] and has much less often called for cost-benefit analysis or open-ended balancing.[16] As Mark Sagoff has correctly argued, this

14. See, e.g., *Union Elec. Co. v. EPA*, 427 U.S. 246, 256 (1976) (the EPA may not disapprove state's implementation of air quality standards based on technological or economic infeasibility); *Weyerhauser Co. v. Costle*, 590 F.2d 1011, 1045 & n.52 (D.C. Cir. 1978) (cost relevant under Clean Water Act only if wholly out of proportion to benefits).

15. The EPA is required to issue such standards governing waste disposal "as may be necessary to protect human health and the environment." 42 U.S.C.A. § 6924(a) (West 1998).

16. The closest approaches to open-ended balancing are found in FIFRA, the Federal Insecticide, Fungicide, and Rodenticide Act, and (somewhat less clearly) in TSCA, the Toxic Substances Control Act. See 7 U.S.C.A. § 136a(a) (West 1998) (EPA regulations are to prevent "unreasonable adverse effects on the environment"); 15 U.S.C.A. § 2605(a) (West 1998) (the EPA is to regulate toxics "to the extent necessary to protect adequately against such risk using the least burdensome requirements").

consistent statutory pattern bespeaks an implicit commitment to environmental rights.[17] Since Sagoff wrote, this pattern has continued with the passage of the 1990 Clean Air Act. Despite a great deal of grumbling from the current Congress, the public still clearly continues to give this baseline overwhelming support. For Sunstein, this overwhelming series of legislative judgments (all contrary to the "neutral" and common law baselines) should provide a powerful basis for an environmentalist baseline.[18]

To make the issue more concrete, suppose the cost of eliminating a carcinogen is just equal to the value of the statistical lives at stake, so a cost-benefit analysis comes out as a tie. In baseball, a tie goes to the runner. What's the rule in the "environmental game"? A libertarian might rule in favor of the polluter on the ground that the company's freedom of action should not be limited without a clear justification. An environmentalist would say that the entitlement to a clean environment should be, at the least, a tiebreaker. What would Sunstein do? At worst, he would flip a coin. At best, having reached this dilemma, Sunstein might recommend that we search for some additional social value that might break the tie. But a priori, he has no more reason to think

17. See Mark Sagoff, *The Economy of the Earth, Philosophy, Law, and the Environment* 179–180, 196–199, 207–208 (1988); Mark Sagoff, "Economic Theory and Environmental Law," 79 *Mich. L. Rev.* 1393, 1396–1400, 1418–1419 (1981).

18. In rejecting this baseline, Sunstein speaks of the inappropriateness of recognizing individual rights with regard to risks that affect huge groups of people. Sunstein, *Rights Revolution, supra* note 12, at 90. The size of the group seems irrelevant, however, except to the extent that he is concerned about possible judicial unmanageability. There is no reason that an entitlement need be held by only a few people. Perhaps the assumption is that these cannot be individual rights because an entire group is indivisibly affected. Id. at 29, 229. Much of contemporary public law litigation involves rights that are inherently shared by groups. See, e.g., Abram Chayes, "The Role of the Judge in Public Law Litigation," 89 *Harv. L. Rev.* 1281, 1291–1292 (1976). In any event, while the language of individual rights may seem strained in the environmental setting, an environmentalist baseline could fit comfortably within a more communal or neorepublican perspective. The right to a decent environment could be seen as resting with the public, as an entity, rather than with individuals. Given Sunstein's leading role in the revival of republicanism, his completely individualistic conception of rights in this context seems a bit puzzling.

that on any given occasion he will adopt environmentalist values (breaking the tie in favor of regulation) than libertarian values (breaking the tie in favor of deregulation).

What is appealing about this proposal is the apparent fairness of having a level playing field, in which neither party enters the dispute with an advantage. This appeal is somewhat ironic in view of Sunstein's attack on the concept of governmental neutrality earlier in his career, but nevertheless it underlies the detachment that pervades his environmental scholarship. But Sunstein was right the first time: neutrality has no privileged status as a value; like any other value it must be considered on its merits in specific contexts. In the environmental context, it is not the value our society has chosen over the course of almost thirty years of vigorous public debate. Since Sunstein does not explicitly identify and defend his chosen baseline, we have to look elsewhere for arguments in its favor.

The Fallacy of Neutrality

The idea of government neutrality is a beguiling one. For example, in a memorable passage, Charles Meyers, then dean of the Stanford Law School, explained the appeal of economic analysis in terms of its purported neutrality: "The environmentalist would base public policy on a set of values he holds to be transcendent and absolute, inherent in the nature of man and therefore ineluctable. The economist rejects absolutes: what is good is what the individual prefers; a good society is one that maximizes freedom of choice."[19] "It is not," Meyers said, that the economist "personally believes that society is better off with a pleasure palace or a power dam at the Grand Canyon; he sees himself as value-free on the question of how the resource is to be used."[20] Similarly, a group of eleven eminent economists, headed by Stanford Nobel laureate Kenneth Arrow, recently endorsed the view that "the values to be assigned to program effects— favorable or unfavorable—should be those of the affected individuals, not the values held by economists, moral philosophers,

19. Charles Meyers, "An Introduction to Environmental Thought: Some Sources and Some Criticisms," 50 *Ind. L.J.* 426, 451–452 (1975).
20. Id. at 450.

environmentalists, or others." Thus, they say, the decision-making process should be a neutral reflection of the trade-offs made in private markets.[21]

Philosopher Ronald Dworkin has provided a fuller defense of this notion of neutrality, based on his general theory of liberalism. Unlike Meyers (but like Sunstein), Dworkin's attachment to "dollar voting" is tempered by concern about equality. Nonetheless, his arguments support Meyers's call for value neutrality on the part of government. Dworkin contends that liberalism requires the government to remain neutral on the question of "the good," taking no position on what forms of human life are better than others. Ideally, the government would give each individual an equal share of resources, leaving it to the market to determine the resulting allocation of goods and services.[22] Thus, if resources are fairly distributed, Dworkin says the government should not promote the views of those who value living in harmony with nature against those who value increased production of consumer goods.[23]

---

21. Kenneth Arrow et al., "Is There a Role for Benefit-Cost Analysis in Environmental, Health, and Safety Regulations?" 272 *Science* 221, 222 (1996). Arrow and his coauthors do leave room for some noneconomic factors such as "equity within and between generations." Id. at 221–222.

22. See Ronald Dworkin, *A Matter of Principle* 191–192 (1985).

23. Id. at 202. Dworkin's approach may be understood within the auction metaphor I have used for cost-benefit analysis. Like the traditional advocates of cost-benefit analysis such as Meyers and Arrow et al., Dworkin does not believe that the government should pursue any values of its own. The values should presumably be those of the affected individuals. See Arrow et al., *supra* note 21. But Dworkin is more worried than Meyers and Arrow et al. about economic equality and objects to their auction technique on the ground that it gives the rich more influence than the poor over the allocation of resources. To understand Dworkin's position, we need to imagine the government giving each citizen vouchers to use in the auction. Each citizen gets an equal amount of purchasing power, so all have the same ability to influence the result. (Thus, the citizens of Duluth won't lose out simply because they're poorer than Reserve's stockholders, as they might under a traditional cost-benefit analysis.) Alternatively, we can imagine a cost-benefit analysis with an egalitarian twist: the WTP measure is not the amount people are actually willing to pay with their current amount of wealth, but the amount they would be willing to pay if current wealth were redistributed equally. Dworkin would argue that this proposal is even more neutral than traditional cost-benefit analysis, since it gives everyone's preferences equal weight in the analysis, rather than weighting those of the wealthy more heavily. Dworkin presents this vision of neutrality as the paragon of liberalism.

As Sagoff and others have argued, however, this view is virtually a caricature of liberalism.[24] While liberals do treasure tolerance and individual freedom, they need not make a fetish of moral neutrality. The primary concerns of liberals should be to ensure that people are not coerced in ways that violate their basic personal freedoms and that society is tolerant of diverse perspectives even when those perspectives are strongly at odds with the majority view. But this does not mean that the majority should always be powerless to implement its own views of public policy. Dworkin's version of liberalism would leave the government only the powers to address issues of rights and inequality—a sort of post–New Deal version of the watchdog state—but would otherwise deny it the power to legislate on any other subject of controversy. That would drain democracy of much of its meaning. Liberalism's dedication toward protecting rights should not be taken to the extreme of strangling the idea of democratic self-rule.

Dworkin does leave the door ajar for environmentalists by raising the possibility that liberals could sometimes support preservation in the name of human diversity, since otherwise some entire ways of life may become extinct and unavailable for future generations.[25] He later makes a somewhat similar suggestion about public support for the arts. There, he says, we should "try to define a rich cultural structure, one that multiplies distinct possibilities or opportunities of value, and count ourselves trustees for protecting the richness of our culture for those who will live their lives in it after us." The goal would not be to give greater pleasure to later generations or to provide a world they will prefer. Instead, the premise is that "it is better for people to have complexity and depth in the forms of life open to them."[26] Surely, a parallel argument could be made for biodiversity.

I will not attempt to develop such an argument for biodiversity here. We can begin to see, however, what the argument would

24. For critiques of Dworkin, see Sagoff, *The Economy of the Earth, supra* note 17, ch. 7; Robin West, "Pragmatism Rediscovered: A Pragmatic Definition of the Liberal Vision," 46 *U. Pitt. L. Rev.* 673 (1985); Steven Shiffrin, "Liberalism, Radicalism and Legal Scholarship," 30 *UCLA L. Rev.* 1103, 1147–1174 (1983).

25. See Dworkin, *supra* note 22, at 202.

26. Id. at 229. In his more recent writings, Dworkin seems somewhat more favorable toward environmental protection. See Ronald Dworkin, *Life's Dominion* 75–76, 154 (1993).

look like by considering descriptions of the significance of bio-diversity. Biologists seem particularly attuned to this value. As one observer explains,

> Since some biologists spend their professional lives sur-rounded by biodiversity, its unfathomable complexity and its sublime beauty combine with feelings of humili-ating ignorance to infuse intense spiritual feelings. The more they learn, the more awe they feel; and the un-knowns, the gaps the sacred world of science can't fill, leave further room for values and spirituality and aes-thetics to rush in.[27]

From "[d]ifference, variety, complexity, heterogeneity, intricacy of individual organisms, organismal interactions, ecological and environmental processes" come the sources of biodiversity's ap-peal to biologists.[28]

In an interview, one biologist remarked that when you look at tiny creatures, you see all kinds of symmetry that might have been created by a great artist and you realize the complexity of the evolutionary process: "That's the mindblowing thing about it. You see it and it's just, God, it's just *beautiful*, absolutely beau-tiful. How did it come about? The process behind it must be even more beautiful, more intricate, more complex, more so-phisticated, whatever."[29] It is this unseen beauty of small things that leads to a leading ant specialist's advice to people who find ants marauding in their kitchens: step carefully; feed them cof-feecake, tuna, and whipped cream—then get a magnifying glass, watch them closely, and "you will be as close as any person may ever come to seeing social life as it might evolve on another planet."[30]

Many of us may favor a different response to domestic intru-sions by insects, but there is something undeniably appealing in this attitude. Most people today recognize that nature has value, quite apart from any immediate utility. Even beyond aesthetic appeal, we can recognize that nature is the result of a process

27. David Takacs, *The Idea of Biodiversity: Philosophies of Paradise* 255 (1996).
28. Id. at 275.
29. Id. at 273.
30. Id. at 310.

beyond human scale, whether in the form of divine intervention or the sheer extent of a billion years of evolution. Together with the more utilitarian reasons for preserving biodiversity to provide direct human benefits, these values deserve a place in our societal pantheon. The government need not be neutral between people holding these values and those who would prefer a sterile world with no organic life apart from humans and their agricultural inventory.

Admittedly, environmentalist values are not universally held. It may not be possible to provide a philosophical refutation of the views of the dissenters, but this does not mean that they are entitled to a veto over the government's decision to favor environmental values. Rather, the government should be entitled to proceed, so long as the values are reasonable (in light of our culture as a whole) and so long as the individual rights of dissenters are respected.

When it endorses moral values, the government is obviously favoring the views of some citizens over the views of others. Anti-environmentalists cannot be happy that their tax dollars are "wasted" on preservation. But citizens often disagree with government decisions and believe that tax money could be better spent elsewhere. The anti-environmentalist taxpayer seems to have no greater grievance than that of the taxpayer who thinks the defense budget is too large. Inevitably, in deciding disputes over public goods, the government will favor some views over others, but this does not mean that some groups are receiving unjustified favoritism. In short, equality doesn't have to mean that government is powerless to favor those who want to save the whales over those who would rather kill them.

In supporting environmental values, the government is not just *arbitrarily* choosing to favor the views of one group of citizens over those of others. Rather, it can make several arguments in favor of the reasonableness of its action. First, this is not an arena (such as religion) in which values can be left solely to private choice. The environment is a public good—we all breathe the same air—so a collective decision is necessary. *Some* choice must be made by the government. Second, environmental values are not only very widely shared, but also deeply embedded in our culture. Wilderness has been an especially important theme in American culture. The Grand Canyon, the bald eagle, and Yellowstone Park are American icons. As we have seen, biodiversity also has connections with the culture of modern science. Even the eco-skeptic is likely to share in one or another of these cultural traditions enough to admit their legitimacy, if not their correctness. Finally, by favoring the environmentalist's perspective, the government is not necessarily treating the eco-skeptic's views as valueless. The environmentalist baseline subordinates, but does not eliminate, nonenvironmental values. In short, the government can properly claim that the environmental baseline is a reasonable resolution of the conflicting values.

To date, neither the advocates of economics like Meyers, nor the neo-republicans like Sunstein, nor the philosophical liberals like Dworkin, have made the case for a neutral environmental baseline. In the previous chapter, I asked how a hypothetical decision maker might go about determining the public interest. Such a decision maker would do well to begin by choosing an environmentalist baseline. Such a baseline is strongly supported by popular "willingness to vote." It can't be rejected on the basis of private preferences because we don't know whether to measure those preferences by willingness to pay or willingness to accept. In thinking about the future of our society, goals like allocational efficiency and distributional equality are all very well. But we can properly hope for more than that, for more even than a society that "takes rights seriously." We can also hope—even if we cannot muster any definitive philosophical proof for our position—for a future society that has preserved natural beauty, has a vibrant culture, and in other ways provides a basis for deeply fulfilling lives.

Up to this point, I've discussed the baseline issue at a fairly abstract level, in terms of contingent valuation measurements, regulatory baselines, and liberal neutrality. Following pragmatist precepts, it may be useful to revert to a more tangible setting. Considering a famous property case may help to bring home the concrete stakes in this dispute. *Boomer v. Atlantic Cement Co.*,[31] a

31. 26 N.Y.2d 219, 257 N.E.2d 870, 309 N.Y.S.2d 312 (1970). My description of the facts is taken from the appellate record. Appendix, Appellant's Court of Appeals Brief, at A53, A65–A66, A67, A70, A77–A78, A102. For further

1970 New York decision, has become one of the great teaching cases in American legal education. It appears in virtually every casebook on the law of property, torts, remedies, environmental regulation, or land use planning. A lucky (or perhaps I should say unlucky) law student can easily encounter the case three or four times. The case provides a striking illustration of the significance of baselines.

Boomer involved nuisance law, which governs a property owner's obligations to its neighbors. When Atlantic Cement opened a new cement plant, the impact on some of its neighbors was drastic. For example, Floyd and Barbara Millious lived in an eight-room ranch-style house nine hundred feet from the cement plant, and even closer to an overhead conveyer used to move rocks to the plant from the nearby quarry. According to the Milliouses, the blasting at the quarry had caused large cracks in the walls, ceiling, and exterior of their home. Moreover, the fine dust blowing onto their property covered everything with what they described as a "plastic-like" coating, which they found impossible to remove.

Kenneth and Delores Livengood lived farther from the cement plant, but closer to the quarry. Delores Livengood's testimony vividly describes the effects of the blasting on their daily lives, as in this episode in March of 1967:

> Well, we were sitting on the floor in the living room, playing a game when they had this awful tremble, and to me the house seemed like the whole house was rocking, and I would see the lamp shades vibrating and it just scared the children. That was one of the times that they just got up and started to run for the basement. And I didn't go out the door because I knew what it was. I heard the big blast, but the whole house just rocked and just wondered if it was going to stop. It was terrible.[32]

The plaintiffs were unquestionably entitled to a remedy of some kind. The question was, What remedy? Noting the size of

---

discussion of the case and its implications, see "Symposium on Nuisance Law; Twenty Years after *Boomer v. Atlantic Cement Co.*," 54 *Albany L. Rev.* 171 (1990).

32. Appendix, Appellant's Court of Appeals Brief, *supra* note 31, at A102.

Atlantic's investment, its use of the best available pollution control, and the size of its work force, the court concluded that closing the plant would be inequitable.[33] Consequently, the court allowed Atlantic to avoid a permanent injunction by paying the plaintiffs "such permanent damages as may be fixed by the court." In return, the plaintiffs would convey a "servitude on the land" to Atlantic.[34] The dissent complained to no avail that the majority was, "in effect, licensing a continuing wrong," thereby allowing the cement plant to seize the property rights of its neighbors.[35]

Maybe this was the right ultimate outcome, given the serious public impact of closing the dominant employer in the area. But the court's opinion invariably disturbs students because of its stance of neutrality between the cement company and the neighbors. Even though the court accepted the finding that the company had committed a tort, its equitable balance seemed to draw no distinction between the wrongdoer and its victims. Under Sunstein's analysis as well, the two sides are treated symmetrically. If, like pollution law, nuisance law is thought to be based on economic efficiency, then according to Sunstein, the dispute would virtually be settled by a cost-benefit analysis; otherwise, the court would have to make a more general balance of the competing public and private interests. Again, what is troublesome is not so much the outcome, but the "neutral" stance of the decision maker, who has no inherent preference one way or the other between the contesting parties. It seems to me, however, that anyone reading the facts cannot help feeling that this is a false symmetry. The Livengoods, hiding from the blasting in the basement with their frightened children, were simply not

33. 26 N.Y.2d at 225, 257 N.E.2d at 873, 309 N.Y.S.2d at 316. For excellent discussions of the "balancing of the equities" in common law nuisance cases, see Dan Dobbs, *Law of Remedies: Damages—Equity—Restitution* 518–528 (1993); Zygmunt Plater, "Statutory Violations and Equitable Discretion," 70 *Cal. L. Rev.* 524, 533–545 (1982); William Rodgers, *Handbook on Environmental Law* 118–121, 344–347 (1977). *City of Aberdeen v. Wellman*, 352 N.W.2d 204 (S.D. 1984), is a modern case rejecting balancing.

34. *Boomer v. Atlantic Cement Co.*, 26 N.Y.2d at 225, 228, 257 N.E.2d at 873, 875, 309 N.Y.S.2d at 316, 319.

35. Id. at 230, 257 N.E.2d at 876–877, 309 N.Y.S.2d at 321 (Jasen, J., dissenting).

on the same moral plane as the company, which had secretly assembled the land for its project and appeared unannounced to destroy their daily lives.

In *Boomer*, there were strong reasons to deny the injunction, which would have been economically devastating to the local community. Use of an environmental baseline would not necessarily have changed the result. The point of the baseline is not simply to control the results of cases, but also to leave us satisfied with the process of reaching the result. The New York courts in *Boomer* actually did not do badly in the end—they left open the cement plant for the benefit of the community, while awarding generous damages that went beyond the dictates of their own analytical framework.[36] But the *Boomer* opinion is disturbing for much the same reason, though to a much lesser degree, as Judge Posner's economic analysis of rape because it adopts a stance of neutrality, rather than beginning with a presumption in favor of the victim. In a more subtle way, the Sunsteinian version of cost-benefit analysis suffers from the same flaw.

For similar reasons, even though Sunstein-style balancing and an environmental baseline might often lead to the same results, the distinction is an important one. To adopt a neutral baseline is not itself a neutral decision; it is based on a value judgment of symmetry between polluters and victims. For thirty years, American public law has rejected that symmetry. Public support for this approach has enabled an environmental baseline to survive, even though for most of the last twenty years, at least one branch of government has been in the hands of its sworn enemies. If we

36. In analyzing *Boomer*, I would favor a baseline of protection against such invasive activities. Rather than leaving the remedy to the unrestrained balancing of the court, I would begin with a presumption in favor of injunctive relief. In this particular case, the presumption might have been overcome by other compelling social interests. Even if no injunction was appropriate, I would argue for using a WTA rather than a WTP measurement of damages in order to uphold the baseline entitlement against outrageous nuisances. Interestingly enough, the lower courts did provide far higher damages than could be justified on the basis of willingness to pay to eliminate the pollution. For example, a restaurant that had never been profitable in the past, and showed no real sign of future profitability, was awarded several hundred thousand dollars for "lost profits." More likely, the compensation was for the destruction of the entitlement to be free from this sort of outrageous invasion of its interests.

believe at all in the idea of public values, adopted as the result of vigorous democratic deliberation, this is a public consensus policy makers must respect.[37]

An Environmental Baseline

It is one thing to endorse an environmentalist baseline and another to work out how such a baseline should apply. Although the baseline is fairly easy to describe in some situations, it turns out to be more difficult in others, as we'll see in the next chapter when we discuss the time dimension of regulation. One important point to make from the start is that the baseline is only that—a baseline, rather than an outcome. The presumption is in favor of environmental protection, but it is a rebuttable presumption. If the choice of baseline provides the "environmentalist" aspect of eco-pragmatism, the methods for moving away from the baseline come from the pragmatist side.

Defining the Baseline

Picking the right baseline involves a complex mix of considerations. I will begin by trying to develop a hybrid of cost-benefit analysis and feasibility analysis, designed to take advantage of the environmentalist tilt of the former and the "reality check" of the latter. I will then examine some objections to feasibility analysis and cost-benefit analysis to ensure that the hybrid approach does not fall prey to the same difficulties.

We can start fleshing out the environmentalist baseline by returning to *Reserve Mining*. The two primary candidates for analyzing such cases are feasibility analysis and cost-benefit analysis. As we have seen, the seemingly drastic differences between these approaches are to some extent deceptive. Under either a feasibility analysis or a risk-averse cost-benefit analysis, the Eighth Cir-

---

37. Strong support for an environmentalist baseline can be found in Donald Hornstein, "Reclaiming Environmental Law: A Normative Critique of Comparative Risk Analysis," 92 *Colum. L. Rev.* 562, 616–620 (1992). See also Christopher Schroeder, Book Review, "In the Regulation of Manmade Carcinogens, If Feasibility Analysis Is the Answer, What Is the Question?" 88 *Mich. L. Rev.* 1483, 1502 (1990); Jay Michaelson, "Rethinking Regulatory Reform: Toxics, Politics, and Ethics," 105 *Yale L.J.* 1891, 1892, 1920–1922 (1996).

cuit arguably reached the right decision in *Reserve Mining*, given the kind of information available at the time. Cost-benefit analysis gives us a more formal framework for weighing costs and benefits, but policy judgments may be decisive when applying the analysis in cases like *Reserve Mining*.[38] Feasibility analysis, on the other hand, purports to make a once-and-for-all policy decision, but the terms "feasibility" and "significant risk" have enough play to allow informal consideration of the cost-benefit relationship.[39] In both analyses, the same factors are probably going to be relevant to the decision, at least in cases like *Reserve Mining*. The results of a cost-benefit analysis by an environmentally sensitive agency may not necessarily differ very much from the results of a sensible feasibility analysis.[40]

Although the differences between these approaches are not as radical as they may at first appear, the distinctions are nonetheless real. Though the two methods of analysis have the potential to lead to the same outcome, they tend to propel the decision maker in different directions. Cost-benefit analysis seems neutral, and the decision maker enters the analysis with no presumption in favor of pollution control. If we assume, however, that the right of the people of Duluth to safe drinking water takes priority over Reserve's right to use Lake Superior as a dumping ground, then this stance of neutrality is wrong-headed. It seems to me that this is the right assumption and that we should there-

38. For example, regulation may almost always be justified if we choose a high enough "value of life" figure. Ultimately, the figures chosen, and thus the decision to regulate or not, depend on certain moral or philosophical beliefs, rather than economic calculations.

39. Chief Justice Rehnquist has recognized the inherent ambiguity of these terms, stating that "Congress required the Secretary to engage in something called 'feasibility analysis.' But these words mean nothing at all. They are a 'legislative mirage' appearing to some . . . but not to others, and assuming any form desired by the beholder." *American Textile Mfrs. Inst. v. Donovan*, 452 U.S. 490, 546 (1981) (Rehnquist, J., dissenting). Chief Justice Rehnquist went on to conclude that the feasibility requirement could be seen to consider a number of variables, including "considerations of administrative or even political feasibility." Id. For this reason, he considered the statute to be an unconstitutionally broad delegation of authority.

40. Similarly, the same regulations OMB now rejects on the basis of cost-benefit, it would also probably reject as "infeasible" or based on an "insignificant risk."

fore begin with a presumption in favor of ending pollution. For this reason, feasibility analysis should be incorporated into the baseline.

The problem with feasibility analysis, however, is that it is either too rigid, if we interpret feasibility to require *all* possible actions short of economic disaster, or too open-ended, if we take it to mean something like "a sensible, practical response, all things considered." It seems to me more appropriate, and more in tune with the overall thrust of federal pollution law, to define feasibility as meaning "not patently disproportional to the potential benefits."

Consequently, I propose a hybrid of feasibility analysis and cost-benefit analysis. I would suggest that we continue to apply feasibility analysis, which Congress has frequently endorsed, but use cost-benefit analysis as a benchmark for what is feasible. When even an environmentally sensitive analysis—using a high value of life, conservative risk estimates, and a low discount rate for future benefits—shows that regulation is clearly unwarranted, we ought to think very carefully about whether a regulation really is a feasible response to a significant risk.

Under this hybrid approach, the proper decision in *Reserve Mining* seems clear. There was by all accounts a potentially serious threat to health; it was technologically and economically possible to eliminate the risk; and the cost-benefit analysis was at least a close call, so that the company could not claim that the costs were patently disproportionate to the benefits. In short, the Eight Circuit had it right, given the information it had to work with.[41]

Defending the Baseline

The early part of this chapter defends the primacy of the environmental baseline and the associated notion of feasibility analy-

41. This combination of cost-benefit analysis and feasibility analysis may strike true believers on both sides as heresy. Aren't there important philosophical issues at stake here? The answer is yes, if you view either method as having some exclusive claim to being the only rational, moral approach to the problem. Even if that should turn out to be so, and I myself am quite dubious on that point, the philosophical difference between the two is of little importance in cases like *Re-*

sis. What remains is to defend the more limited role played by cost-benefit analysis. The attacks can come from two directions: environmentalists may fear the malign effect of economics on our environmental values, while economists may wonder why, having admitted the nose of the cost-benefit camel, we shouldn't let the beast into the tent entirely. I will begin with the environmentalist concern that *any* use of cost-benefit analysis will subvert environmental law.

At the heart of much of the opposition to cost-benefit analysis is a sense that economics may undermine valuable social norms and impoverish social discourse. Cost-benefit analysis is often attacked for treating human lives as commodities. I agree that the language used by economists can be dehumanizing.[42] Critics are deeply concerned about the "expressive function" of law and the way in which economic discourse may warp that expressive function.[43] As two of these critics recently put it, "[C]hoices among alternative approaches to law and policy making—especially the choice between cost-benefit and other approaches—are significant apart from the results they produce." Such choices "reflect *how* we think about various social 'goods'—and how we think is a matter of independent ethical significance because it helps define the kind of community we will have."[44]

According to such critics, economics celebrates the marketplace, but markets can be "isolating, competitive, mean-spirited

---

serve Mining. If *Reserve Mining* is at all typical of current issues in environmental law, we might do well to leave the theoretical arguments aside and to focus our own energies on more practical concerns.

42. See Clayton Gillette and Thomas Hopkins, *Federal Agency Valuations of Human Life: A Report to the Administrative Conference of the United States* 25 (1988). Compare Sagoff, *The Economy of the Earth, supra* note 17 (arguing that issues of human health and environmental quality are fundamental ethical considerations that cannot be reduced to economic variables) with Carol Rose, "Environmental Faust Succumbs to Temptations of Economic Mephistopheles, or, Value by Any Other Name Is Preference," 87 *Mich. L. Rev.* 1631 (1989) (book review) (arguing that ethical values are not so readily distinguishable from economic preferences and that the use of market rhetoric forces consideration of environmental issues by individuals who are not otherwise inclined to do so).

43. Sunstein, *Free Markets, supra* note 1, at 91–93.

44. Jane B. Baron and Jeffrey L. Dunoff, "Against Market Rationality: Moral Critiques of Economic Analysis in Legal Theory," 17 *Cardozo L. Rev.* 431, 433 (1996).

arenas."[45] As Ron Cass, dean of Boston University Law School, rephrases the objection, the analogy is to the doctor who refers to patients by disease rather than name ("the ruptured disc in Room 102") and no longer thinks of his patients as real people: "For him, the problem no longer is how George Johnson feels and functions but what mechanical treatment seems necessary to a particular organ or muscle or other sub-part. Such language may be appropriate to the efficient discussion of a technical problem, but it has a long-term, subtle, corrosive effect."[46]

This is a legitimate point. A society in which people thought about themselves and others purely in the language of economic theory would be a repulsive spectacle. It would be appalling if we lost the capacity to use any other language than that of economics to describe our social world—or to discuss environmental issues, for that matter. But this is a far cry from the rather modest use of cost-benefit analysis and other quantitative methods in my proposal. By placing the primary emphasis on the environmentalist baseline, on the contrary, I would hope to reinforce the law's ability to express environmentalist values.

While we should not forget the expressive function of law, we should not allow that function to blind us to its instrumental significance. It would be morally bankrupt to focus solely on beautifying the language of the law while remaining heedless of the direct consequences of legal decisions. Cass's essay was prompted by the debate over Judge Posner's famous (or infamous) proposal to use market mechanisms to deal with a shortage of adoptable babies. Cass thoughtfully canvassed the objections to using the depersonalized language of economics in this context. He closed, however, with an observation about how the debate over depersonalization may sacrifice the individuals most affected to abstract disputes over language and technique. Because so much of the debate has focused on symbolic issues, he said, "there should be at least a pang of regret" at losing sight of more concrete consequences: "For people whose lives are vitally

45. Id. at 495.
46. Ronald A. Cass, "Coping with Life, Law, and Markets: A Comment on Posner and the Law-and-Economics Debate," 67 *B. U. L. Rev.* 73, 76–77 (1987). Cass ultimately rejects this objection.

affected by legal and institutional constraints on the adoption process, a straightforward evaluation of what truly is gained or lost by *this* proposal is of immense importance."[47] From the pragmatist's perspective, Cass is right about the need to keep instrumental values in sight. It is for this reason that we need cost-benefit analysis to ensure that we do not allow our commitment to environmental ideals to turn into fanaticism.

If cost-benefit analysis is attacked for cold-bloodedness, feasibility analysis is often considered soft-headed and wedded to foolishly expensive methods of controlling environmental problems. Feasibility analysis is often attacked for requiring inefficient "command and control" regulations, in which the EPA directs particular firms to achieve the specific level of pollution control it considers feasible.[48] The hybrid approach might be implemented through conventional regulations, but we could adopt a different strategy instead: use the hybrid approach at the industry or even global level to determine the appropriate level of environmental quality; then use marketable permits and other nonregulatory measures to implement that level of control. By cautioning against wastefully inefficient methods of regulation, the hybrid approach may help nudge the government in this direction. Thus, it would be a mistake for economists to link the hybrid approach with disfavored methods of "command and control" regulation. Given the wide band of uncertainty around environmental cost-benefit analyses, it would also be a mistake to assume that a stronger role for formal cost-benefit analysis would necessarily lead to substantially greater economic efficiency in practice.

Besides its substantive merits, from a pragmatist point of view, the hybrid approach is appealing in terms of regulatory process. Both feasibility analysis and cost-benefit analysis have some positive aspects in terms of institutional performance. Cost-benefit analysis can be more easily standardized across agencies and across regulatory decisions. If we are worried about the quality of agency decision making, then OMB and the courts can exercise control more effectively when agencies use a formalized pro-

47. Id. at 97.
48. Sunstein, *Rights Revolution, supra* note 12, at 87–88.

cedure like cost-benefit analysis. The main advantage of feasibility analysis is precisely its lack of formality, so it can be used more comfortably where the data is messier and the social values are less easily quantified.

Even though both tools are potentially useful, there are serious pitfalls in relying exclusively on cost-benefit analysis, which the hybrid approach avoids. Despite its positive qualities, using cost-benefit analysis to control decisions (rather than as a source of information) can warp the administrative process. One of the risks of cost-benefit analysis is that it may obscure policy choices behind seemingly technical decisions (such as the choice of a discount rate). Another risk may be that the locus of decision making is shifted toward OMB and away from agencies like the EPA. In the view of some observers, at least, this may result in what amounts to crude political pressure from the White House, masked by economic jargon.

There may also be a more subtle harm in shifting decisions away from officials who are really knowledgeable about the substance of the decision to bean counters at OMB. Cost-benefit analysis reflects a very traditional type of organizational structure, in which underlings collect and analyze numerical data that provides the basis for a superior's decision. We might learn from the private-sector experience with similar management techniques.[49]

Cost-benefit analysis came to Washington with Bob McNamara and his "Whiz Kids" at the Defense Department.[50]

---

49. Rather than being considered the equivalent of the finance department, OMB might be considered the counterpart of the factory inspector. Using that analogy, however, OMB still can be criticized as an example of the outmoded concept of inspecting quality into products. See E. Donald Elliott, "TQMing OMB: Or Why Regulatory Review under Executive Order 12291 Works So Poorly and What President Clinton Can Do about It," *L. & Contemp. Probs.*, Spring 1994, at 167. Both Elliott's analysis and mine invite the response that business management techniques are irrelevant to government. An extended rebuttal to that criticism is provided by David Osborne and Ted Gaebler, *Reinventing Government: How the Entrepreneurial Spirit Is Transforming the Public Sector* (1992). It is nevertheless important to keep in mind that the government's substantive goals are likely to be quite different from those of the private-sector firm and that the government must also consider equity and process values as well as efficiency.

50. See Susan Rose-Ackerman, *Rethinking the Progressive Agenda: The Reform of the American Regulatory State* 15 (1991).

Some evidence suggests that one problem of American industry in the past few decades—and of McNamara's own legacy at Ford Motor Co. in particular—was a shift in corporate influence from manufacturing and sales executives to the finance department.[51] Applying techniques like cost-benefit analysis, financial analysts rejected bold automotive innovations because the potential benefits could not be adequately documented.[52] While it would surely be an overstatement to assign the primary blame to corporate bean counters for the struggles of some American industries, the best decisions are more likely to be made by people immersed in the substance of the business, rather than in accounting. Instead, many companies increasingly vested power in individuals whose expertise was in economics and finance theory, but who had little contact with the product or the operation of the business.[53]

The Reagan and Bush administrations made a similar effort to turn operations over to the regulatory accountants. Thus, it seems, the federal government may have been attempting to replicate an organizational mistake that had already contributed to the decline of key American industries.[54] Today advocates of the Contract with America seemingly wanted to recreate the same

51. See David Halberstam, *The Reckoning* 213–216, 245–246 (1986). For example, Ford's European branch developed an improved method to paint cars in 1958, but the method was not fully adopted at Ford's American plants until 1984, id. at 499–501, because of the influence of the "finance men," since there was no way to quantify the substantially improved quality in terms of sales, id. at 500. It's hard to overlook the obvious analogy to disputes between environmental regulators, who claim a new regulation will advance environmental quality, and OMB, which objects that the benefits cannot be proven greater than the costs.

52. Consider the following description of the "tyranny of the number crunchers" at GM:

> Another form of financial tyranny has been the frequent decision to pinch pennies at the expense of product content and design. Cars have often been produced that lack the exciting features once planned because original designs have had to be altered so many times to save money. Of course, the long-term results of penny-wise-and-pound-foolish car-building practices are legendary. . . .

Maryann Keller, *Rude Awakening: The Rise, Fall, and Struggle for Recovery of General Motors* 28 (1989).

53. See Halberstam, *supra* note 51, at 313–315.

54. OMB's lack of substantive expertise has been a frequent basis of criticism. See Thomas McGarity, *Reinventing Rationality* 281 (1991).

problem on a grander scale. We ought to resist suggestions that we continue farther down this path.

Apart from leading to a more workable administrative structure than would exclusive reliance on cost-benefit analysis, the hybrid approach has a more important virtue. However we make environmental decisions, value choices are necessary. In cost-benefit analysis, the value choices are hidden in technical decisions about discount rates, valuation problems, and the like. It is difficult for the outsider—not just the ordinary voter, but also legislators, journalists, and other interested observers—to understand the meaning of arcane factors such as the discount rate, let alone to distinguish between purely technical and value-laden considerations in setting these parameters. The concepts used in the hybrid approach—significant risk, feasibility, and gross disproportionality—are far more understandable to outsiders and put the value judgments out on the table where they can be seen. In a democracy, this is no inconsiderable virtue.

A devotee of cost-benefit analysis might question why, having admitted that this approach is a legitimate part of the decision process, we should give it only limited scope, rather than making it the central mode of analysis. The answer I have tried to develop here has three parts. First, as a practical matter, I doubt that cost-benefit analysis is capable of giving us the kind of clear-cut answers that would be needed for it to play this role. Trying to make cost-benefit analysis decisive would demand a degree of precision that the analysis cannot yet supply. Second, as a matter of institutional design, relying wholly on cost-benefit analysis would leave difficult policy choices in the hands of the wrong people. Economic analysts can make a useful contribution, but they should not be responsible for the ultimate policy choices, being neither democratically accountable nor expert in the substance of environmental issues. Finally, it seems to me, relying strictly on cost-benefit analysis would not do justice to our community's values and would to some degree trivialize our national commitment to the environment. Normative decisions cannot ultimately be reduced to a unidimensional balance, and our political discourse would suffer from an effort to force environmental law into such a reductionist mode as cost-benefit analysis.

In short, cost-benefit analysis should assist rather than control regulatory decisions. It functions best as a critical resource to prevent misguided decisions, rather than as an effort to make hard social decisions on spreadsheets. Indeed, if past experience is any guide, that is probably the most OMB can realistically hope to accomplish anyway.[55] The hybrid approach, then, gives cost-benefit analysis just the right scope: enough to serve as a check on unreasonable regulation, but not enough to take over the decision process.

The Hybrid Approach and the Courts

Not surprisingly, law professors devote much of their attention to the development of legal doctrine by the courts. My primary focus in this book, however, is on the trade-offs between environmental and economic values made by various decision makers—sometimes courts, but more often agencies and legislatures. Nevertheless, the hybrid baseline also has some important implications for how judges should approach issues of legal doctrine.

Once again, *Reserve Mining* provides a useful illustration. Much of our discussion in *Reserve Mining* focused on the policy question facing the court, which was how to weigh the potential for harmful health effects against the costs of land disposal. But the court also faced some significant doctrinal questions relating to methods of statutory interpretation and to judicial discretion in enforcing statutes. Although space does not allow a full exploration of the implications of the hybrid baseline for legal doctrine, a brief discussion of these issues may provide at least a glimpse of what a fuller treatment would look like.

The Greening of Environmental Statutes

The first doctrinal issue involved statutory interpretation. The *Reserve Mining* court was faced with the need to interpret some ambiguous language in the Clean Water Act. The statute allowed the government to seek judicial relief when discharges "endanger . . . the health or welfare of persons." The issue of statutory interpretation was whether the existence of a *potential*

---

55. See Kip Viscusi, "Equivalent Frames of Reference for Judging Risk Regulation Policies," 3 *N.Y.U. Envtl. L.J.* 431, 450 (1994).

threat to public health met this standard of endangerment. Calling for a "common sense" interpretation of the statute, the court concluded that the term should be construed in a "precautionary or preventive sense," rather than requiring clear proof of a significant risk.[56] Application of the hybrid approach might have provided a firmer basis for this conclusion, as I will argue below.

In interpreting statutes today, courts apply a number of "canons" of interpretation: for instance, that waivers of sovereign immunity must be express and that ambiguous criminal statutes are construed in favor of the defendant.[57] But there is no particular canon dealing with environmental issues. The hybrid approach would suggest interpreting ambiguous statutes to cover significant environmental risks (with an escape hatch for infeasibility). Should such a "green" canon be recognized?

Statutory interpretation, as a topic, is the subject of vigorous scholarly and judicial debate. In general, there seem to be three basic positions current in the literature. First, the conventional view is that judges should attempt to apply statutes in accordance with the legislative intent. (The whole concept of legislative intent is itself subject to vigorous attack, based largely on skepticism about whether groups of legislators share a coherent, let alone public-spirited, set of intentions. Nevertheless, this approach still has its adherents.) Second, advocates of dynamic interpretation believe that judges should apply evolving public values when interpreting statutes. Advocates of this perspective also call on courts to use their rulings to improve the deliberative aspects of the legislative process. Third, formalists such as Justice Scalia argue that interpretation should be based not on current values or legislative intentions, but on the language actually enacted by the legislature, construed in light of general rules and canons of interpretation.[58] Although the question is obviously a

56. *Reserve Mining Co. v. United States,* 514 F.2d 492, 527–529 (8th Cir. 1974).

57. For a good explanation of the purposes served by such canons, see Daniel Rodriguez, "The Presumption of Reviewability: A Study in Canonical Construction and Its Consequences," 45 *Vand. L. Rev.* 743, 747–751 (1992).

58. Rather than citing directly to the extensive scholarship on these subjects, it seems more useful to refer the reader to the best collection of materials on statutory interpretation, William Eskridge and Philip Frickey, Cases and Materi-

complex one, under each of these three views there is an argument to be made in favor of a green canon of interpretation.

The argument is straightforward in terms of the "legislative intent" approach. One justification for canons of interpretation is that they mirror the likely intentions of the legislature. Courts assume that the legislature would have been explicit if it had wanted to deviate sharply from well-established legal principles. After all, legislators are likely to share in the general acceptance of those principles, at least to the extent that a decision to deviate would be hotly debated. Even critics who believe that Congress has been unduly responsive to environmental concerns, despite their normative qualms, are endorsing the empirical proposition that this is indeed what Congress often intended. Thus, given the frequency with which Congress has applied some form of the hybrid approach, it seems plausible to assume that this was the legislative intent in a given case even if the language used is somewhat ambiguous.

In terms of dynamic interpretation, the argument is also straightforward, at least if we assume that judges are supposed to look to evolving community norms rather than their own personal preferences when construing statutes. For, as I have argued, the environmentalist baseline is clear and well established. As the community's considered judgment, reached after considerable political deliberation, this norm is well deserving of judicial respect. Moreover, forcing Congress to be explicit when it wishes to depart from the prevailing baseline has the advantage of making it more politically accountable and less prone to behind-the-scenes manipulation by private interests.

Perhaps more surprisingly, there is also a formalist argument for a green canon. The argument is grounded on the National Environmental Policy Act (NEPA), the statute that mandates environmental impact statements. Although the statute is best known for that mandate, it also contains other provisions, including a broad statement of environmental policy. NEPA calls on the government to use "all practicable means, consistent with

---

als on Legislation: Statutes and the Creation of Public Policy (2d ed. 1995). My own views are sketched in Daniel Farber and Philip Frickey, *Law and Public Choice: A Critical Introduction* 88–115 (1991).

other essential considerations of national policy," to achieve a list
of environmental goals. These goals include directives to "fulfill
the responsibilities of each generation as trustee of the environment for succeeding generations" and to "assure for all Americans safe, healthful, productive, and esthetically and culturally
pleasing surroundings." The statute also declares the "critical
importance of restoring and maintaining environmental quality
to the overall welfare and development of man."[59] The Supreme
Court has ruled that a court has no power to review whether a
particular agency action comports with these policies, assuming
a valid impact statement exists.[60] Nevertheless, there is a strong
argument in favor of applying these policies to the interpretation
of ambiguous statutes.[61]

For a formalist, statutory text is the decisive consideration.
NEPA itself expressly establishes a canon of statutory interpretation. Section 102(1) contains the following unmistakable mandate: "The Congress authorizes and directs that, to the fullest
extent possible (1) the policies, regulations and public laws of
the United States shall be interpreted and administered in accordance with the policies set forth in this chapter."[62] If this language is not clear enough to establish a canon of interpretation,
it is hard to imagine what would be required to do so. Note that
this mandate has two parts—it endorses environmentalist goals,
but subjects them to a qualification ("to the fullest extent pos-

59. NEPA § 101, 42 U.S.C.A. § 4331 (West 1998).

60. *Robertson v. Methow Valley Citizens Council,* 490 U.S. 332 (1989).

61. The one escape I can see for the formalist would be a claim that interpretative directives by Congress violate the separation of powers by instructing
judges how to decide cases. See Alan Romero, "Interpretive Direction in Statutes," 31 *Harv. J. on Legis.* 211 (1994). This argument strikes me as unconvinc-
ing in the context of NEPA, where Congress is neither decreeing a methodology
for interpreting statutes (such as requiring use of legislative history) nor dictating
results in particular cases, but merely creating national policy, as it might do with
a "purpose" clause in any individual statutory provision. Cf. *Robertson v. Seattle
Audubon Soc'y,* 503 U.S. 429 (1992) (interpretative directive constituted change
in law rather than effort to "direct any particular findings of fact or applications
of law, old or new, to fact" and was therefore constitutional).

62. 42 U.S.C.A. § 4332(1) (West 1998). For another call on courts to attend
to this language, see Nicholas Yost, "NEPA's Promise Partially Fulfilled," 20
Envtl. L. 533, 539 (1990).

sible"). This language seems virtually identical with the hybrid approach that I have endorsed in this chapter.

Thus, regardless of whether we think judges should implement legislative intent, evolving community values, or textual commands, a strong case can be made for a "green" canon of interpretation.[63] Statutory interpretation is a knotty subject, and the issues cannot be fully resolved in the course of a few paragraphs. But what I have presented here will, I hope, be enough to put the possibility of an environmental canon of interpretation on the table.

Making Judicial Discretion Environmentally Friendly

Another doctrinal issue in *Reserve Mining* related to the scope of judicial discretion. Traditionally, courts have had broad discretion in issuing injunctions. In the *Boomer* case, discussed earlier in this chapter, the state court balanced the interests of the company and the neighbors in order to arrive at an equitable result. Similarly, in *Reserve Mining,* the court attempt to "strike a proper balance between the benefits conferred and the hazards created by Reserve's facility."[64]

The Supreme Court has addressed the issue of equitable discretion in two significant cases, but has created as much confusion as enlightenment on the issue. *TVA v. Hill,*[65] better known as the *Snail Darter Case,* involved a nearly completed dam. The Court held that it had no choice but to enjoin completion of the dam, thereby wasting the $100 million already spent on the project, because completing the dam would have destroyed an endangered species of fish. Four years later, however, in *Weinberger v. Romero-Barcelo,*[66] the Court allowed the Navy to continue military exercises without having obtained a permit required by the Clean Water Act. The Court viewed the harm to water quality as minimal, the violation as purely technical, and

63. See also Sunstein, *Rights Revolution, supra* note 12, at 184 (passing remark expressing approval for a canon of interpretation in favor of protection of "noncommodity" environmental values).
64. 514 F.2d at 535.
65. 437 U.S. 153 (1978).
66. 456 U.S. 305 (1982).

the governmental interest in proceeding as strong. The *Weinberger* Court distinguished the *Snail Darter Case* as involving an
absolutist statute and instead stressed the tradition of broad equitable discretion in the issuance of injunctions.[67]

Putting aside for the moment the specifics of the particular
statutes involved, the hybrid approach would generally counsel
against either open-ended balancing or absolutism in environmental injunctions. Instead, as a general rule, the court should
enjoin an environmental violation unless compliance would not
be feasible—that is, would involve costs grossly disproportionate
to any environmental benefit. *Weinberger* fits easily under this
rule, since the Court found no environmental harm and serious
governmental costs. The outcome in *Reserve Mining* also fits the
hybrid approach, as we have seen earlier in this chapter.

The *Snail Darter Case* requires more careful analysis. An honest cost-benefit analysis would have shown—as a subsequent
government study proved—that the dam was unjustified even in
economic terms, never mind the environment.[68] So the actual
outcome of the case is easily squared with the hybrid approach:
the environmentally desirable outcome was not only economically feasible, but also economically desirable. Still, the absolutist language of the opinion is clearly a deviation from the approach that I suggested here. Was such an absolutist approach
justified?

This question needs to be addressed at two levels. First, we
must ask whether the Court itself properly eschewed any consideration of cost in its decision. The answer here seems to me

67. For further discussion of these cases, see Daniel Farber, "Equitable Discretion, Legal Duties, and Environmental Injunctions," 45 *U. Pitt. L. Rev.* 513
(1984). The Court also touched on the issue in a later case, *Amoco Production Co.
v. Village of Gambell, Alaska*, 480 U.S. 531 (1987), where it refused a preliminary
injunction against a procedural violation of a statute dealing with Alaskan lands.
The Court did observe that if environmental injury is "sufficiently likely," "the
balance of harms will usually favor the issuance of an injunction to protect the
environment."

68. For a retrospective view of the case, see Zygmunt Plater, "The Embattled
Social Utilities of the Endangered Species Act: A Noah Presumption and Caution against Putting Gasmasks on the Canaries in the Coalmine," 27 *Envtl. L.*
845 (1997).

straightforward. The hybrid approach is only a baseline, and Congress had clearly directed decision makers to take a more absolutist approach in deciding individual cases under the Endangered Species Act. Within constitutional limits, this is the kind of choice that legislatures are entitled to make and that courts must respect even when the legislatures have chosen unwisely. But this merely pushes the question to a second level, whether Congress was justified in precluding decision makers from considering cost.

At first blush, this absolutist congressional mandate may seem inconsistent with the hybrid approach. In another sense, however, the Endangered Species Act can be considered an application of the hybrid approach, but on a categorical rather than a case-by-case basis. Congress obviously felt that destruction of endangered species posed a threat of serious environmental harm, and it also seems clear that compliance by the government with the statute *overall* was feasible—that is, there is no reason to think that avoiding the destruction of endangered species, as a general matter, involves costs that are grossly disproportionate to any environmental benefit. Thus, at a "macro" level, the statute is consistent with the hybrid approach.

The tricky problem is whether to apply the hybrid approach at this macro, across-the-board level, as in the original Endangered Species Act, as opposed to the micro, case-specific level of *Reserve Mining*. In general, individual tailoring seems preferable to "one size fits all," but some strong arguments exist for taking a more sweeping approach with respect to endangered species. It is easier to assess the general benefits of saving endangered species than to assess the value of any one species, and there is a serious risk that decision makers will be tempted to sacrifice the less ascertainable value of a species to the pressing imperative of completing a favored project. So the refusal of Congress to allow application of the hybrid approach to individual cases may not have been unreasonable, given that the statute as a whole probably does fit the hybrid approach. Yet completely ignoring all countervailing considerations seems excessive. After *TVA v. Hill*, Congress responded by providing a special administrative mechanism for cases in which the social cost of preserving a species is unacceptably high, while still refusing to countenance consideration of any cost on a routine basis.[69] As Zygmunt Plater puts it,

> given the array of good but subtle reasons for protecting endangered species, public values that are typically not readily marketable and that provoke a bitter marketplace backlash, we should adopt a Noah Presumption, a strong presumption in favor of protecting all endangered species . . . unless human necessities clearly outweigh the importance of doing so.[70]

As the history of the Endangered Species Act illustrates, knowing how to apply the hybrid approach can involve some subtle institutional issues. It is not always obvious where in the regulatory system the comparison between costs and benefits should be made. Originally, Congress made the determination itself on a purely categorical basis; more recently, it has provided a special administrative mechanism for doing so. As the Court correctly concluded, however, Congress has not called on judges to make this determination in issuing remedies for violation of this statute.

TVA v. Hill demonstrates that it would be wrong for courts to apply the hybrid approach without regard to the specific statute that they are implementing. Nevertheless, I would argue, we could bring clarity to the problem of enforcement injunctions by recognizing a general presumption in favor of enjoining violations of environmental statutes. The presumption could be overcome by showing that the costs of compliance are grossly disproportionate to any environmental benefit. Of course, it is always open to Congress to mandate the use of a different standard, as it did in the Endangered Species Act.

This discussion of legal doctrine has taken us a bit away from the general focus in this chapter on policies concerning environmental risks. But the detour seems justified, if only to indi-

69. Congress created a special cabinet-level committee to determine whether the proposed action's benefits to the public clearly outweigh the benefits of compliance and whether there are reasonable alternatives to the action. 16 U.S.C.A. § 1536(e), (h) (West 1998). This retreat from absolutism, while guarded, fits the general pattern of movement toward the hybrid approach.

70. Plater, *supra* note 68, at 847.

cate that the approach adopted here has broader applications. A full exploration of the doctrinal implications of the environmental baseline might be the subject of another book, but this brief discussion may be enough at least to show the hybrid approach's potential for illuminating doctrinal issues.

RETURNING TO THE ISSUE of risk management, it may be useful to take stock. In a nutshell, I have advocated the following principle for dealing with risks, at least where those risks can be roughly quantified: "To the extent feasible without incurring costs grossly disproportionate to any benefit, the government should eliminate significant environmental risks."

The primary virtue of this hybrid approach is that it corresponds to society's recognition of a presumptive entitlement to environmental quality, but without losing sight of the costs of environmental protection. By using cost-benefit analysis as a constraint on regulation, it avoids asking more of the cost-benefit analyst than she can legitimately deliver, but it does take advantage of her ability to toss a dose of cold water on overheated environmental sentiments. Institutionally, it places the primary responsibility for regulation with the line agencies whose policy analysts are most experienced with specific problems, but it leaves room for OMB to play a healthy role as a critic of regulation. General principles do not resolve hard cases, but this hybrid approach at least provides a framework for analysis.

The environmentalist baseline is not intended to be a radical departure from current regulation. On the contrary, the same basic concept appears in statute after statute, whether under the rubric of feasibility or some variant of the BAT requirement. But recognizing the environmental baseline explicitly would allow us to clarify two aspects of current practice. First, Congress sometimes purports to impose absolute standards of environmental quality, but then backslides by allowing time extensions or variances when compliance proves impossible. Rather than slipping the notion of feasibility through the legislative backdoor in this fashion, we ought to recognize that feasibility is an integral condition of the environmental mandate. Second, in a few situations, a regulation is feasible, but involves trivial benefits at high cost. Often, current statutes do not explicitly allow the EPA to take this disproportionality into account, which may lead the EPA to "fudge" on its statutory directions or provoke a wasteful dispute with OMB or Congress. The suggested approach would allow more straightforward and sensible decisions in these situations. Unlike current proposals for regulatory reform in Congress, however, this approach would continue to recognize the environmentalist baseline and would limit cost-benefit analysis to a supporting role.

The remainder of the book is devoted to charting some other dimensions of the environmentalist baseline. By assuming rough quantifiability of risks, I have passed over the great uncertainty surrounding these estimates. What should we do when the risks seem serious, but we don't know enough to begin to quantify them? I have also ignored the special difficulties of defining "substantial risk" and "disproportionate cost" when these take place in different time periods, perhaps involving different generations. Is spending $1 million today in order to save a statistical life forty year later "disproportionate"? Finally, what I have offered so far is an essentially static analysis: what to do if you must decide, on a fixed and incomplete record, the extent of regulation to impose for all time. In reality, environmental regulation is a continuing process, not a one-shot affair. Much of the debate over environmental policy suffers from having too static an orientation, but I will try to provide at least some outlines of a more dynamic approach to using an environmental baseline.

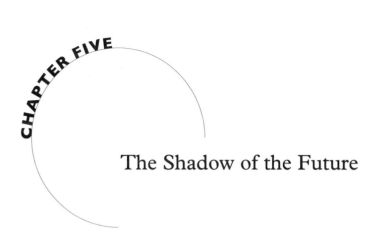

The Shadow of the Future

Should we spend $10 million to save a single life? This is a classic problem of comparing apples and oranges. If that question isn't sufficiently challenging, suppose the money must be spent today, but the life won't be saved for a century. Now we're being asked to compare today's apples with oranges from trees that haven't yet been planted.

One reason environmental decisions can be so difficult is that they extend deep into time. We simply are not used to making decisions in ordinary life with such long time horizons. But the problem is not simply the sheer passage of time. It is compounded by our ignorance of environmental problems and by the rapid evolution of scientific knowledge, so that what we think we know today is likely to need correction tomorrow. The dangers that seem most urgent today may fade in the light of greater knowledge, while other dangers may turn out to be more severe than we expected. Chapters 5 and 6 consider how environmental pragmatism can respond to these challenges.

In this chapter, we will put to the side the problem of changing scientific knowledge and focus on how to factor time lags into decisions. This is a problem regardless of how we approach environmental trade-offs. Suppose we base our regulatory decision solely on the presence of a significant risk and decide that an annual risk of one chance in a million is significant. But what if this risk is far in the future? Is a risk that is spread over a fifty-year time period as significant as one that occurs tomorrow? Before answering "yes" too readily, we might reflect that lengthy time periods can inflate a minuscule risk to overwhelming proportions. For example, a one-in-a-trillion risk will only cause a death in the United States once every few centuries. But over eternity, this adds up to an infinite number of deaths, presumably justifying an infinite expense by the current generation.

Time lags are also a problem for cost-benefit analysts. Suppose you are a cost-benefit analyst faced with the following question: If it is worth spending $1 million to save a life today, how much should we spend to save a life in twenty years? During the Reagan and Bush administrations, OMB would have answered $150,000.[1] Before 1992, OMB used a 10 percent discount rate to convert future regulatory costs and benefits into their "present value."[2] Since 1992, OMB has used a 7 percent rate, which would change the answer to $260,000 (over 70 percent higher than the earlier figure).[3]

To understand discounting, consider the work of a retirement planner. You tell the planner how much money you want to have a certain number of years in the future. The planner checks on available interest rates and tells you how much money you need to put aside today. The higher the interest rate and the longer the time period, the less you need to invest today to have a given amount in the end.

Like looking into the wrong end of a telescope, discounting makes distant objects look smaller. Because government regulation of carcinogens won't affect the cancer rate for twenty or thirty years,[4] OMB's choice of discount rates had dramatic implications for regulatory policy. Just as with the retirement plan, a higher interest rate reduces the amount today that is equivalent

1. OMB's recent reduction of the regulatory discount rate to 7 percent (see *infra* note 3) would make regulation more attractive.
2. OMB Circular A-94 at 4 (1972).
3. OMB Circular A-94 (1992 revision).
4. The EPA estimates the "latency" period for asbestos-related lung cancer at twenty years from the time of exposure and the latency period for asbestos-related mesothelioma at twenty-five to thirty years. *Final Rule: Asbestos Manufacture, Importation, Processing and Distribution in Commerce Prohibitions*, 54 Fed. Reg. 29,460, 29,469 (1989).

to a given amount in the future. Compound interest can have remarkable effects over long periods of time.

In *Reserve Mining*, as we saw in chapter 3, the discount rate was a key factor in determining whether land disposal was cost-justified. Because the costs of land disposal would be paid in just a few years, whereas any decline in cancer rates would not be felt for decades, how to treat the time lag was a crucial question. But the time lag problem is not simply a technical economic issue. In addressing the trade-offs in *Reserve Mining*, we don't need to use the specific methodology of cost-benefit analysis. Instead, we could ask the following question: If society is willing to spend $5 million to prevent a statistical death tomorrow (based on whatever approach to environmental values we favor), how much should we be willing to spend today to prevent a death twenty years from now?

To stake out the boundaries of the debate, consider two purist positions. On the one hand, we might reject discounting entirely, so the $5 million figure continues to be the test for whether Reserve should be forced to clean up. On the other hand, we might try to take advantage of the growth of investments over time. Reserve could put aside a smaller amount of money now, reinvest the proceeds, and easily have enough twenty years later to spend $5 million on other safety measures that would save equal numbers of lives. Given Reserve's 10 percent rate of return on assets, it would need to set aside only around $750,000 today in order to have $5 million in twenty years. Depending on which argument we accept, we would find that it is worth spending either the full amount today or only one-seventh as much, due to the time lag. This is, to say the least, a fairly significant issue in setting regulatory policy.

The effect of discounting on the value of a regulation's costs or benefits can vary dramatically, depending on the size of the discount rate and the delay before costs or benefits are realized. Because the benefits of environmental regulation often come long after the costs,[5] compound discounting generally has a

5. See Clayton Gillette and Thomas Hopkins, *Federal Agency Valuations of Human Life: A Report to the Administrative Conference of the United States* 54–57 (1988). This is true in toxics regulation due to the long latency period between exposure

greater impact on the present value of benefits.[6] A few simple numerical examples illustrate how dramatically a high discount rate erodes present value.

Consider how the discounted present value of a million dollars, to be received in twenty years, falls with different discount rates. It is $1 million with a zero discount rate, $380,000 at 5 percent, $150,000 at 10 percent, and only $60,000 at 15 percent. Over a forty-year period, the present value of $1 million nearly vanishes when calculated at a 15 percent rate.[7] (Suppose someone offered you a government bond that paid $1 million in forty years. If you can get 15 percent on your other investments, you should be willing to swap the million dollar bond for an old car—the right to receive a million dollars that far in the future is only worth a few thousand today.) Given these dramatic figures, it should be no surprise that methods of discounting are critical to cost-benefit analysis and often pivotal in regulatory decisions.[8]

---

and the onset of toxic-related disease. For example, virtually all the costs of the EPA asbestos rule would have accrued within six years. 54 Fed. Reg. 29,461–29,462. The EPA estimated the benefits of the regulation over thirteen years, and the Fifth Circuit criticized the agency's failure to estimate the benefits further into the future. *Corrosion Proof Fittings v. EPA,* 947 F.2d 1201, 1218–1219 (5th Cir. 1991).

6. "Present value" is the term used to describe the current value to the recipient of a benefit to be conferred at some time in the future. For example, suppose Linda agrees to loan Barb $100 in return for a payment of $121 in two years, charging 10 percent simple compound interest. The present value to Linda of $121 in year 2 is $100. The 10 percent annual rate Linda uses to discount the money she will receive in year 2 is called the "discount rate."

Future monetary costs may be discounted to present value in a similar manner. If financial benefits to the same consumer and decreases in consumption were not discounted at the same rate, odd results would occur. For example, in a future "wash transaction," the costs and benefits will cancel when the transaction occurs, giving the transaction no net economic effect. But the cost and benefit would discount to different numbers, so the present value of a future wash transaction would not be zero.

7. Compound rates from Peter G. Sassone and William A. Shaffer, *Cost-Benefit Analysis: A Handbook* 128 (1978).

8. "Experience has shown that cost-benefit results are generally quite sensitive to the discount rate." Id. at 128. See Ann Fisher, "An Overview and Evaluation of EPA's Guidelines for Conducting Regulatory Impact Analysis," in *Environmental Policy under Reagan's Executive Order: The Role of Cost-Benefit Analysis* 99, 108–111 (V. Kerry Smith ed., 1984) (commenting that selection of the discount

The discount rate has the greatest impact on long-term global environmental issues such as the ozone layer and the greenhouse effect.[9] For instance, using OMB's old discount rate of 10 percent, if the greenhouse effect were to cost society $100 billion twenty years from now, it would be worth spending only $11 billion today to prevent the harm. Under OMB's new discount rate of 7 percent,[10] it would be worth spending a more generous $25 billion.

Most people know the fable of the future-oriented grasshopper and the present-oriented ant. From an economist's point of view, the difference between them can be easily quantified. The grasshopper may know perfectly well that winter is coming, but has a high discount rate, which eliminates any incentive to save for the future. The ant, on the other hand, has a low discount rate and is willing to defer gratification now for rewards in the future. In terms of the fable, OMB's shift in discount rates is a move from a grasshopper-like approach to a more ant-like attitude toward the future.

When they are first introduced to the concept, discounting seems to most people like the ultimate in "policy wonk" technicalities. What earthly difference does it make whether we choose to use 5, 7, or 9 percent in cost-benefit calculations? As we've just seen, the practical answer is that it makes a tremendous difference. Switching from a 10 to a 7 percent rate seems like a small adjustment, but it might mean doubling our investment in preventing global warming. What is ultimately at stake is how heavily we weigh the future against the present. On the spectrum between grasshoppers, who live only for today, and ants, who

---

rate is critical to determination of the result of EPA regulatory impact analysis under Executive Order No. 12,291).

9. See Maureen Cropper and Wallace Oates, "Environmental Economics: A Survey," 30 *J. Econ. Literature* 675, 725–727 (1992) (noting the significance of choice of discount rate in the analysis of carcinogen regulation under CERCLA and RCRA (Resource Conservation and Recovery Act) as well as for global environmental issues).

10. Regulatory Working Group, *Economic Analysis of Federal Regulations under Executive Order 12866* (Jan. 11, 1996) <http://www.whitehouse.gov/WH/EOP/OMB/html/miscdoc/riaguide.html>.

work only for tomorrow, where should our society be? Answering this question involves profound normative questions as well as technical economic analysis.

Depending on the time span, discounting may or may not be the best method for addressing the fundamental question of future orientation versus present-mindedness. When we are dealing with a relatively short period of time such as twenty years, costs and benefits come during the lifetimes of members of a single generation. The people who finance the pollution improvements today are mostly the same people as those who will hopefully experience a lower cancer rate in twenty years. What we need to know is how the delay affects the way they value the trade-off between cost and benefit. Discounting can be a useful way to approach the problem, as we will see in the next section.

Over longer time spans, discounting is less appealing as a methodology. Even small discount rates begin to have startlingly large effects, so that future effects seem to be undercounted. For example, at a 7 percent rate, well over 99 percent of a disaster in 2098 would vanish when shrunk to present value. Perhaps more important, it is no longer true that the costs and benefits are experienced by more or less the same people, so an important question of fairness arises. What obligation do we have to make sacrifices for the benefit of our great-great-grandchildren? That question will be considered in the final section of the chapter.

Discounting within Lifetimes: Grasshoppers versus Ants

Because present value calculations are so sensitive to the choice of interest rate, using the correct discount rate is critical for cost-benefit analysts. Finding the correct discount rate requires a deeper analysis of why people prefer a given quantity of present benefit over the same quantity of future benefit. Economists emphasize two explanations: money could be invested for a greater future return (the opportunity cost of capital), and people are impatient (time preference).[11] In a ideal world of fully competi-

11. My discussion loosely follows the description of social opportunity cost and social time preference in David W. Pearce and R. Kerry Turner, *Economics of Natural Resources and the Environment* 212–217 (1990). Most cost-benefit texts contain similar discussions. Time preference could also reflect the expectation

tive markets, no taxes, perfect information, and complete rationality, the rates would be identical. Reality is not quite so tidy. Most economists agree that the discount rate suggested by the impatience explanation—the "social" discount rate—is substantially lower than the rate indicated by the opportunity cost of alternative investments.[12] Current estimates for the social discount rate, based on the long-term real rate of return on riskless investments (Treasury notes and bonds), are in the neighborhood of 1 percent. The discount rate based on opportunity costs is significantly higher.

The Social Discount Rate

One rationale for discounting is a simple preference for receiving a benefit today over receiving the same benefit tomorrow. As a matter of human psychology, the observation that people are often impatient is manifestly correct. However, it is debatable whether impatience and preferences based on that emotion are rational or a prudent basis for public policy decisions.[13] After all,

---

that future societies are likely to be richer, making an extra dollar of benefit less valuable in the future than it is to the current society. This rationale for discounting is often called the "diminishing marginal utility" argument.

12. In a world without taxes, the social discount rate and the opportunity cost theoretically should be the same. But the tax system drives a wedge between the two. See Robert C. Lind, "A Primer on the Major Issues Relating to the Discount Rate for Evaluating National Energy Options," in *Discounting for Time and Risk in Energy Policy* 21, 24–32 (Robert C. Lind et al. eds., 1982). For example, if individuals use a 2 percent discount rate for personal consumption, they will be willing to save only if given at least a 2 percent return. To generate a 2 percent return after taxes to consumers, firms must invest in projects offering a higher return. If business and personal taxes take a combined "bite" of 50 percent out of firm income by the time it reaches shareholders, the firm will need to earn twice as much in order to give shareholders their 2 percent after-tax return. Thus, in this simple example, the social discount rate is 2 percent, while the implicit opportunity cost of capital is 4 percent. This distinction between the social discount rate and the opportunity cost of capital turns out to have crucial importance for cost-benefit analysis.

13. Rawls asserts that, assuming a future benefit is no less certain than a current benefit, preferring one simply because it is closer in time is irrational. John Rawls, *A Theory of Justice* § 45 (1971). I am not fully convinced, however, that future and present experiences rationally *should* be weighted exactly the same. To demand that our present selves give the same weight to future experiences as to present ones seems in some sense to deny the reality of time, asking us to

one of the lessons we teach children is to delay gratification because we know that tomorrow does eventually come. Time preference is the attitude of the grasshopper in the fable, not that of the more farsighted ant who prepares for the inevitable winter.

Trying to calculate the right discount rates involves difficulties not unlike determining the economic value of life. Besides the moral issues raised by the entire exercise, there are complex technical issues. As with the value of life, different studies come up with different values, partly because methodological difficulties exist, but also because individuals' behavior may not be entirely consistent. There is also room to doubt whether the figures reflect fully informed, voluntary choices by the individuals involved. It's worth spending a little time to explore these technical issues.

According to economic theory, rational individuals should use a single discount rate for both saving and borrowing over all time periods. The empirical evidence is quite different. The "real" (inflation-adjusted) rate of return on riskless investments is quite low, approximately 1 percent or so.[14] On the other hand, people are willing to borrow money at much higher rates, even while keeping low-interest investments.[15] They also seem to discount

---

treat time as simply an illusion that should be ignored for purposes of rational decision making.

14. See Barbara Fried, "Fairness and the Consumption Tax," 44 *Stan. L. Rev.* 961, 985–986 (1992) (most reliable estimate is that real, riskless rate of return is 0.5 percent); Robert C. Lind, "Reassessing the Government's Discount Rate Policy in Light of New Theory and Data in a World Economy with a High Degree of Capital Mobility," 18 *J. Envtl. Econ. & Mgmt.* S-8, S-24 (1990) (government's borrowing cost is 1–3 percent); Charles H. Howes, "Introduction: The Social Discount Rate," 18 *J. Envtl. Econ. & Mgmt.* S-1, S-2 (2 percent estimate); Lind, "A Primer on the Major Issues," *supra* note 12, at 73, 83–84 (real risk-free rate of return "near zero").

15. See George Loewenstein and Richard H. Thaler, "Intertemporal Choice," *J. Econ. Persp.*, Fall 1989, at 181, 181. See also Deborah M. Weiss, "Paternalistic Pension Policy: Psychological Evidence and Economic Theory," 58 *U. Chi. L. Rev.* 1275, 1300–1311 (describing differences in a given individual's time preferences, depending on temporal distance and psychological factors, neither is adequately explained by a consistent time-preference theory); Lind, "Reassessing," *supra* note 14, at S-19 to S-21 (discussing consumer borrowing and investing at inconsistent rates and compartmentalization of investment decisions, resulting in a single individual exhibiting several disparate rates of time preference).

future gains differently than future losses, contrary to conventional economic theory.

Sometimes people will even pay money in order to save, as in the once-popular Christmas clubs. These clubs offered the opportunity to lock up funds with no interest (meaning a real loss of value, given inflation) so the money would be available during the holiday season. A more contemporary example is presented by people who deliberately overwithhold on their taxes so the IRS will collect savings on their behalf during the course of the tax year, which they will receive back (with no interest) as a refund after filing their returns.

At least some disparities involve people's desire to commit themselves to various levels of savings, which may make it rational to tie up some funds for a 2 percent return while borrowing on a credit card at a much higher real rate. As Robert Lind explains, "[I]t may not be irrational for the individual to keep budgets separated because of problems associated with self-control," since the "person who regularly raided the children's college fund to pay off consumer debts might soon find that the children had no money for college." For similar reasons, many people "deliberately don't take more than a certain amount of money to Las Vegas or to a race track."[16]

The empirical evidence about environmental discount rates is also conflicting. A survey by economists at Resources for the Future asked a thousand Maryland households about their views regarding saving human lives. On average, people discounted future lives saved within twenty-five years at an annual rate of 8.6 percent, but used an annual rate of 3.4 percent if the time horizon is a century.[17] A Swedish study found much lower rates for longer time horizons, in the neighborhood of 0.0001 percent.[18]

16. Lind, "Reassessing," *supra* note 14, at S-19. This insight may have important implications regarding the appropriate discount rate for very long-term projects.

17. "What Price Posterity?" *Economist,* Mar. 23, 1991, at 73, 73. Neither the time preference theory nor the opportunity cost theory provides any explanation for discounting benefits to be received further in the future at a lower rate than more immediate benefits.

18. Maureen L. Cropper and Paul R. Portney, "Discounting and the Evaluation of Lifesaving Programs," 3 *J. Risk & Uncertainty* 369, 375 (1990).

Responses to such surveys are remarkably varied. In one study, about 10 percent of the respondents had *negative* discount rates.[19] In other words, they counted future harms more heavily than current ones. Many others had in effect infinite discount rates: they refused to give *any* weight to deaths occurring many years in the future. Their responses were based partly on general uncertainty about the future and partly on the belief that science would surely discover a method of eliminating any risk in the meantime.[20] Adding to the confusion, an effort to determine how much people discount their own lives in the future came up with a rate of about 2 percent, close to the return on riskless investment.[21]

It is hard to know what to make of all this. Economic theory assumes a degree of consistency regarding time preferences that seems questionable in the real world. There are also genuine normative concerns about this kind of discounting. Nevertheless, it seems to me that a small discount rate in the neighborhood of 1 or 2 percent is appropriate.

To begin with, the discount rate used for government decisions clearly should not be *higher* than the real rate of return on riskless investments. There is a broad consensus among both economists and the public that American savings rates are actually too low.[22] The judgment that savings rates are "too low"

19. Maureen L. Cropper et al., "Discounting Human Lives," 1991 *Am. J. Agric. Econ.* 1410, 1412. In another study, one-third of the respondents used zero or negative discount rates. John K. Horowitz and Richard T. Carson, "Discounting Statistical Lives," 3 *J. Risk & Uncertainty* 403, 410 (1990). This implies that the weight those respondents assign to future lives is equal to or greater than the weight they accord present lives.

20. Cropper et al., *supra* note 19, at 1412, 1415.

21. Michael J. Moore and W. Kip Viscusi, "Discounting Environmental Health Risks: New Evidence and Policy Implications," 18 *J. Envtl. Econ. & Mgmt.* S-51, S-59 to S-62 (1990).

22. Charles Schultze, *Memos to the President: A Guide through Macroeconomics for the Busy Policymaker* 236–254 (1992); Robert C. Lind, "The Rate of Discount and the Application of Social Benefit-Cost Analysis in the Context of Energy Policy Decisions," in *Discounting for Time and Risk in Energy Policy* 443, 445 (Robert C. Lind et al. eds., 1982); Janet L. Yellen, "Symposium on the Budget Deficit," 3 *J. Econ. Persp.*, Spring 1989, at 17, 17. It could be argued that the reason savings are too low is that people's actual discount rates are above the current riskless rate of return. But it seems to me that a more plausible explanation is that people are too shortsighted to save enough to optimize their future con-

means that people are not being sufficiently future-minded in their saving and borrowing behavior, which, in turn, means that they are discounting the future too heavily.[23] If so, the government should use a lower rate than people are already using for private investment decisions.[24]

Thus, the social discount rate should be no larger than the rate of return on riskless investments. Should it be lower? Although the question probably has more theoretical than practical significance, it is not easy to resolve. There is substantial appeal to the idea of a zero rate, since a death today and a death tomorrow are in some fundamental sense equal. But it is important to recall that here we are dealing only with discounting within a particular generation, not with obligations to later generations.[25] This means that the same individuals are involved in both relevant

---

sumption flows, either because they are too impulsive to limit private borrowing or because political problems create an incentive for government deficits.

23. "'We are living well by running up our debt and selling off our assets. America has thrown itself a party and billed the tab to the future. The costs, which are only beginning to come due, will include a lower standard of living for individual Americans and reduced American influence and importance in world affairs.'" Yellen, *supra* note 22, at 17 (quoting economist Benjamin Friedman); see B. Douglas Bernheim, "A Neoclassical Perspective on Budget Deficits," *J. Econ. Persp.*, Spring 1989, at 55, 55 (public opinion polls showing federal budget deficits as critical national economic concern, second only to unemployment). For some possible reservations about the existence or significance of a shortfall in savings, however, see Daniel Shaviro, *Do Deficits Matter?* 58–59, 168–170, 211–220 (1997).

24. Furthermore, as we have seen, empirical studies show that people use a variety of discount rates in different situations. Among these rates, there is a good argument that the return on riskless investments is the most relevant. Unlike some of the empirical studies of how people would make hypothetical choices, investment rates reflect actual decisions and therefore are a more accurate indicator of preferences. As compared with many borrowing rates (such as those on consumer credit), investments are less likely to be impulsive and more likely to reflect thoughtful deliberation. They are also more likely to reflect long-term preferences, as opposed to short-term desires for cash or the practical unavailability of certain goods such as homes except on credit. Finally, individuals seem protective toward their long-term investment strategies, even when this requires rather expensive efforts to protect those strategies against short-term impatience. This suggests that their considered judgments about time preferences are better reflected by investment returns than by interest rates on consumer debt.

25. Thus, we can put to the side the arguments against intergenerational discounting made in Tyler Cowen and Derek Parfit, "Against the Social Discount Rate," in *Justice between Age Groups* 144 (Peter Laslett and James S. Fishkin, eds., 1996).

time periods. The question is whether, in considering costs or benefits to a particular individual, the government should apply a lower discount rate than that individual herself applies in reasonably well considered personal decisions.[26] Such a policy would raise concerns about paternalism, which at least puts the burden of proof on the proponents of a zero rate.

Thus, the riskless investment rate should be used as both a ceiling and a floor for the social discount rate. According to the most recent empirical evidence, this translates into a discount rate of roughly 1 percent. Accordingly, if we discount future deaths for purposes of cost-benefit analysis, we should use a very low rate. I will suggest later, however, that we may be able to dodge entirely the need to discount future deaths, even within an economic framework.

Losing Investments While Saving the Environment

So far, we have been talking about impatience as a reason for discounting. Another justification for discounting future regulatory benefits is that dollars invested to comply with regulation

26. Setting the discount rate at zero would also leave it a bit below the rate of return on riskless investments (such as government bonds), which also supplies the discount rate for ordinary consumption. This disparity creates the possibility of paradoxical results, in which it becomes optimum for society to precommit to future regulations even though at no point in time is actually implementing the regulations ever considered to be worth the immediate cost. It seems perverse that society should precommit to adopting a regulation that society finds unwarranted today and will find equally unwarranted when it finally goes into effect.

An example may clarify this point. Suppose society believes that it is worth spending $5 million to eliminate a statistical death from a given hazard (but no more), that a certain regulatory requirement could save a life at a cost of $6 million, and that neither of these figures changes over time. Then at no time will society ever want to be subject to the regulation. But society would agree to the following scheme: today place $4.5 million in a trust fund, investing only in risk-free securities; the trustee is to be legally obligated, when the investment compounds to $6 million, to spend it to implement the regulation. The present value of the future life saved is $5 million (because we are applying a zero discount rate), while the present value of the regulatory cost is $4.5 million. Thus, the trust fund scheme is justified by a cost-benefit analysis and should be adopted although it commits society to expenditures that will exceed the benefits when they occur.

The same logic would support a similar device for *any* regulation, no matter how expensive, sufficiently far in the future. In effect, this scheme places a higher

could have been invested elsewhere. Because the benefits of the investment (saved lives) are not realized for several years, society "loses" the interest it would have obtained on those dollars if they had been earning interest elsewhere. Discounting accounts for these forgone investment opportunities.

This rationale applies to the *cost* of regulatory compliance, not to the value of regulatory *benefits*. Just because business investments earn interest does not mean that a life saved today earns interest at the same rate to become two lives twenty years from now. Conversely, if a regulation saves two lives twenty years from today, it makes little sense to say that the opportunity cost of saving those lives means that those two future lives are identical in intrinsic value to one life today. Similarly, it is an open question whether full lung capacity and ability to breathe freely at age thirty are any less valuable than at age twenty.[27] But even if we don't discount the future benefits of regulation, we must somehow take into account the economic reality that the cost of environmental regulation partly takes the form of diminished investment for other purposes.

Traditionally, the loss of alternate investment opportunities has been taken into account by using the rate of return on alternate investments as the rate for discounting benefits. This actually provides a measure of opportunity cost only if the lost opportunity actually *is* an investment whose returns accrue in the same year.[28] Thus, before using this approach, we need to know the time profile of the alternate investment. We also need to consider the possibility that this alternate investment either would never be made at all or would be made even if we go ahead with the regulation. It is hard to know how much expendi-

---

practical significance on saving lives the further in the future they are, but when the day actually arrives, society always regrets having made the investment.

27. See Miley W. Merkhofer, *Decision Science and Social Risk Management* 101 (1987). Viscusi's answer to this argument is that it asks the wrong question. He argues that even assuming an equal willingness to pay to avoid such a loss at the time of occurrence, the money an individual is willing to pay at age twenty should be given greater weight because that money could be invested and would thus have a higher "terminal value." W. Kip Viscusi, "Valuation of Risks to Life and Health: Guidelines for Policy Analysis," in *Benefits Assessment: The State of the Art* 196 (Judith D. Bentkover et al. eds., 1986).

28. See Lind, "A Primer on the Major Issues," *supra* note 12, at 50–52.

tures for government regulations "crowd out" alternative investments.

Instead of this traditional approach, economists have increasingly endorsed a different way of handling opportunity costs by using a "shadow price" for capital. The idea is to trace the future returns (including reinvestments) that are lost if some capital has been diverted by a government action. In other words, the opportunity cost is expressed as a flow of returns to consumers from alternate investments. The shadow price analysis then converts this flow to present value using the social discount rate.[29] For instance, suppose the long-term return on investment is 7 percent and the investment in regulatory compliance displaces an equal amount of other private investments. Then to measure opportunity costs, we would spin out annual returns to investors at the 7 percent rate and use a 1 percent discount rate to reduce those future returns to present value.

When the subject of a cost-benefit analysis is a government project like a new dam, determining the alternate investment return can be difficult. It may depend on whether the project is funded through taxes or increased debt; whether government debt is sold abroad or domestically; the time profile of alternate private investments; the propensity of consumers to save; and the extent to which private savings go toward covering depreciation, rather than new investment. Because the results are sensitive to initial assumptions, analysts have some concerns about the practical feasibility of using this method of discounting for government-financed projects.[30]

Some economists argue that the problem is usually simpler

29. See id. at 41–42, 50; Joel D. Scheraga, "Perspectives on Government Discounting Policies," 18 *J. Envtl. Econ. & Mgmt.* S-65, S-69 to S-70 (1990). The seminal work was David Bradford, "Constraints on Government Investment Opportunities and the Choice of Discount Rate," 65 *Am. Econ. Rev.* 887 (1975). For a concise statement, see David A. Starrett, *Foundations of Public Economics* 194 (1988).

30. See generally "Symposium: The Social Discount Rate," 18 *J. Envtl. Econ. & Mgmt.* S-1 to S-71 (symposium on current state of theory and practice of discounting costs and benefits of government regulation); see also Lind, "Reassessing," *supra* note 14.

for government regulations (rather than government expenditures).[31] Because private investments are normally written off through depreciation, capital is not permanently withdrawn from the investment pool.[32] So we need to account only for the lag between the initial investment and the depreciation recovery. Alternatively, we need consider only the interest payments that are passed on to consumers. Essentially, the question is much like that of determining whether a new toll road is economically viable: we need to compare the benefits received by consumers to the amortized capital costs.[33]

This technique is still relatively new and untested. OMB describes the approach as theoretically preferable, but highly sensitive to assumptions. Some of the relevant assumptions involve access to global capital markets; government control over aggregate investment; and relations among domestic saving, investment, and regulatory costs. As a result, agencies wishing to use this method are cautioned to consult OMB.[34] The technicalities of determining the shadow price of capital are beyond the scope of this book. What is important is that the rate of return on private investments is important only for determining opportunity *costs*. It is not logically relevant to determining the discount rate for regulatory *benefits*. Those should be discounted (if at all) through use of the social discount rate, which is normally much lower.

31. For a summary, see A. Myrick Freeman, *The Measurement of Environmental and Resource Values* 213–217 (1993).

32. This assumes costs are passed on to consumers. (For a discussion of the situation in which costs are paid solely from retained earnings, see Lind, "The Rate of Discount," *supra* note 22, at 450–451.) The details of the simplified version of the shadow price theory can be found in Jeffrey Kolb and Joel Scheraga, "Discounting the Benefits and Costs of Environmental Regulations," 9 *J. Pol'y Analysis & Mgmt.* 381 (1990). Their approach translates into the equivalent of a conventional discount rate of 5–10 percent, depending mostly on the "tail" over which benefits occur. Because they used a 3 percent estimate for the consumption rate of interest, while I suspect a figure closer to 1 percent would be a better estimate, I believe that these figures may need adjustment. See id. at 388.

33. Cf. Richard A. Epstein, "Justice across the Generations," 67 *Tex. L. Rev.* 1465, 1483 (discussing financing a toll road and implications for intergenerational equity).

34. Regulatory Working Group, *supra* note 10.

Beyond Discounting

Accounting for the opportunity cost of investments seems sensible and morally unobjectionable. Discounting future benefits is more troublesome, even given a low discount rate. It's hard to deny a certain discomfort with the idea of discounting future deaths. But it's also hard to deny that there's an "apples and oranges" problem with comparing future benefits with present costs. As we've seen, one way to put both sides into comparable units is to discount the future benefits to present value. This has become the conventional economic technique, perhaps because it's like the way businesses make investment decisions. But there is an alternative that would overcome the "apples and oranges" problem without the moral discomfort of discounting future lives. Rather than creating comparability by discounting future benefits to the present, we could achieve comparability by projecting present costs into the future. The mathematics is essentially equivalent: all we need to do is identify an appropriate future flow of costs whose discounted present value is the same as the actual current cost.

For instance, suppose we are trying to decide whether to spend $10 million today to clean up a waste site. The cleanup will save no lives for twenty years and then will save one life per year forever, starting in 2020. If the shadow cost of capital is 7 percent (to use the current OMB discount rate), then $10 million today is equivalent to about $40 million payable in twenty years. (As usual, the tricky point is picking the right discount rate. There aren't many investments that lock up capital for this long, and some of the return on shorter-term investments may leak away as consumption, rather than being reinvested.) But the deaths do not occur all at once, so we need to project the $40 million in a steady stream forward from 2020. Given the 7 percent shadow cost of capital, this translates into an annual expenditure of $2.8 million beginning in 2020.

So far, what we've done is to convert a current expense of $10 million into an equivalent annual stream of $2.8 million payments commencing twenty years from today. But now we have an "apples to apples" comparison within each individual future year: one life saved for the equivalent of a $2.8 million expenditure in each year after 2020. Is this worthwhile? The answer is yes if we expect people in those years to place a $2.8 million value or more on saving a statistical life. Notice that we've managed to overcome the problem of intertemporal comparisons *without* discounting any lives.

Because all we've done is to reconfigure the math, this maneuver doesn't change the results of any decisions. Nor does it solve the knotty problem of establishing the right discount rate. What we have accomplished, however, is to avoid the nasty connotations of treating future deaths as less significant than present ones.

Fairness toward Our Descendants

In dealing with carcinogens, we are addressing risks that will materialize within a single generation: the people who are now exposed are usually the ones at risk for future harm. But some major environmental problems have much longer "tails." Global warming, for example, can be expected to have an impact over centuries, rather than decades. In *Reserve Mining,* the health effects were probably limited to the existing population using the water supply, since the fibers were unlikely to remain circulating in the lake indefinitely. It's easy to imagine, however, a situation in which the effect on the lake would have been more permanent—for instance, if a proposed nuclear storage site created a risk of releasing radioactive materials into Superior. These very long-term effects present something of a puzzle for policy analysts who must decide how to weigh them against current costs.

Obviously, long-term effects shouldn't be ignored. Yet if we don't somehow take into account the time lag, even relatively trivial effects can balloon. For example, given the risk of being hit by lightning (about 1 in 1,000,000), it would be worthwhile for the U.S. government to spend about $500 million to eliminate the risk completely in any particular year, given a $2 million value of life. If we count future deaths as heavily as present deaths, that means we should be willing to spend an infinite amount—or at least the entire gross domestic product—if doing so would allow us to permanently eliminate the risk of anyone ever being struck by lightening. Or if you prefer not to put dollar values on human life, think of it this way: if we ignore time lapses, we should be willing to sacrifice a hundred thousand lives today (allowing the complete destruction of the population of Duluth, for example) if doing so would eliminate the threat of any American ever being struck by lightning for a millennium (a total of about 240,000 lives). This is, to say the least, counterintuitive. So it seems that we need to somehow take the time lag into account. The problem is that discounting seems to be the wrong technique for doing so.

Discounting and Future Generations

Discounting favors regulations with benefits in the present or near future over regulations with later benefits. One might even say that the *purpose* of discounting is to favor present benefits over future benefits. Discounting also generally favors regulations that produce short-term benefits and long-term costs.[35] Even a modest discount rate favors small benefits today over much larger benefits in the distant future.

For instance, assume we are considering two proposals for nuclear waste disposal.[36] The first option is to build a repository that will last for five hundred years, but will almost certainly leak radiation and cause one billion deaths in the year 2500. Also suppose we know that no lives will be lost in the construction of the repository proposed in option one. The second option, because of special construction hazards inherent in its design, will likely result in the death of at least one worker, but probably no more than two workers, in the year of construction. The second option, however, will *not* leak in five hundred years and thus will not cause any deaths in the year 2500. Finally, assume a 5 percent discount rate.

Although unhappy with the choice, most people would probably prefer to save one billion future lives at the cost of one current life, even though the benefit would be conferred on society

35. See Gillette and Hopkins, *supra* note 5, at 62 (citing Peter Railton, "The Search for A Single Metric" (1987) (unpublished manuscript)).

36. For the sake of simplicity, also assume that the monetary costs of the two proposals are equal and that construction of either repository would be completed in a single year.

five centuries in the future. So they would pick the second option. However, if a policy maker uses cost-benefit analysis with discounted benefits, she will choose the first option because the one billion lives in the year 2500 have a lower present value than one life today.[37] Implicitly, she is valuing a *billion* future lives as being worth less than *one* current life.

This hypothetical vividly illustrates how discounting may discriminate against future generations. Even in more realistic scenarios, discounting often reduces large benefits in the far future to present insignificance. Our previous discussion concerned regulations whose benefits and costs accrue primarily to the same generation. When a government regulatory decision confers costs or benefits on future generations, additional issues arise.

The most compelling questions regarding long-term effects pertain to the rights of future generations and the present generation's obligations to the future. Most people do agree we have at least some responsibility to consider and provide for the welfare of future generations.[38] As mentioned in chapter 4, NEPA

37. This hypothetical is built on a discussion in Derek Parfit, *Reasons and Persons* 357 (1984).

38. Discussions of these issues include "Agora: What Obligation Does Our Generation Owe to the Next? An Approach to Global Environmental Responsibility," 84 *Am. J. Int'l L.* 190–212 (1990) (symposium); James Woodward, "The Non-identity Problem," 96 *Ethics* 804 (1986); Parfit, *supra* note 37 at 480; Brian Barry, "Justice between Generations," in *Law, Morality and Society* 268 (P. M. S. Hacker and J. Raz eds., 1977); Rawls, *supra* note 13, at 44–45 (1971). Much of the debate turns on the problem of how to assess moral duties toward individuals whose very existence depends on our own actions. Adoption of any major social or regulatory program will have at least some effect on people's lives, and hence on whom they marry, the dates of their children's birth and so forth, all of which means that children will be born who are not identical to those who would have lived without the program. If someone leads a happy life, but dies at age forty because of the program, is she worse off than she would have been without the program, if otherwise she would never have been born? If not, can the program violate any moral obligation owed to such members of future generations? Even minor programs may have substantial effects on the composition of future generations, due to what is called the "butterfly effect" in chaos theory. Anthony D'Amato, "Do We Owe a Duty to Future Generations to Preserve the Global Environment?" 84 *Am J. Int'l L.* 190, 192–193 (1990). Where major programs and long time-spans are involved, the identity of virtually every member of a future generation may be changed by our decisions. Although the philosophical issues are intriguing, I agree with Derek Parfit (who first raised the whole prob-

explicitly recognizes such intergenerational duties regarding the environment. Proceeding on that assumption, one facet of the debate about discounting focuses on what discount rate, if any, is consistent with our responsibility toward future generations.

Some of the problems can be addressed through the "projection" technique discussed in the previous section. We can convert the current investment into a future economic return so we can directly compare lives saved in the future with the equivalent costs during the same period. If we are trying to decide whether it is worth investing a million dollars today to save a hundred lives a century from now, we can begin by asking whether people, either then or now, would be willing to sacrifice the compounded value of the million dollars in order to save that many lives. As before, this avoids having to discount the value of the future lives to present value. Of course, the problem of determining the right interest rate to use for such a long-term investment may be difficult. But the toughest problem is that the costs of the regulation are expended by today's generation, while the benefits will be incurred by the as-yet-unborn future generation. This raises a profoundly difficult question of fairness.[39]

The Puzzle of Equity between Generations

Here is a partial list of vexing problems raised by intergenerational effects:

- Is it meaningful to speak about the rights of people who do not currently exist and whose very existence may depend on the decisions whose effects we are trying to assess?

---

lem of duties to "potential persons") that they have little practical relevance. See Derek Parfit, "Comments," 96 *Ethics* 832, 855 (1986).

39. Suppose it turns out that the forward-projected future cost of the regulation, using the appropriate interest rate, turns out to be $200 million, or $2 million per life saved. We can expect that both we and our descendants would probably view this as an appropriate trade-off for saving a current life. This doesn't completely settle the issue, however. Just because our descendants would be willing to spend $2 million to save one of their own lives, it does not necessarily follow that we today are morally obligated to sacrifice the same amount of *our* money to benefit other people who do not yet exist.

- Should the interests of all generations count equally, or do less remote generations count for more?
- Should we assume that our descendants will be richer than ourselves (and therefore undeserving of wealth transfers) or poorer (and therefore in a position to complain about our refusal to share the wealth)?
- Would it be better to have a future with a large number of people and a relatively low (but decent) standard of living or to have fewer, but wealthier, people?

I don't pretend to have answers to these profound questions.[40] As with some of the other problems we have discussed, however, it may be more useful to focus on practical environmental choices, rather than abstract conundrums. As a practical matter, regulation cannot feasibly stray too far from the deeply held values of the public. Environmental regulations must be realistic about sacrifices for future generations. Whatever may be true in the abstract about our duties to future generations, we know that people are willing to make some sacrifices for their descendants, but only within limits. Any practical scheme of environmental protection must function within those limits. We can usefully consider what a practical scheme of long-term environmental planning would have to look like in order to remain effective over multiple generations.

Because we are dealing with such long-term issues, even a decision maker who gives no weight to current public opinion must be concerned with the public's future views. To be sustainable, a long-term environmental program must be capable of maintaining public support over the long haul. Otherwise, the program cannot hope to survive long enough to be effective. Of course, decision makers may sometimes gamble on the future, hoping that public opinion will shift in their direction. I doubt, however, that most people are likely to have a radical change of heart about such fundamental questions as their own responsibilities toward their children and grandchildren. At the least, it seems foolhardy to place environmental policy on such a shaky

---

40. For an exploration of the philosophical issues, see Laslett and Fishkin, *supra* note 25.

footing as a gamble on fundamental future changes in personal values.

As a practical matter, it is also unlikely that members of the current generation will be willing to make greater sacrifices for anonymous members of future generations than they are willing to make for their own immediate descendants. Thus, feelings of obligations toward one's own descendants provide an upward *practical* limit on obligations toward future generations as a whole.[41] Hence, accepted ideas about responsibilities toward descendants provide a useful benchmark for duties to future generations generally.

This benchmark enables us to invoke some widely shared intuitions. First, it is not clear whether the language of "moral obligation" is appropriate when considering unborn descendants. If your great-grandparents squandered the family fortune, you may feel that they acted reprehensibly, but it would be difficult to charge them with violating a personal obligation toward you or with violating your "rights." For this reason, I think the language of "responsibility" rather than "obligation" is more appropriate: mature individuals behave responsibly with respect to the interests of their descendants, but do not necessarily owe a duty to as-yet-nonexistent individuals. The current generation is morally constrained by the interests of future generations, but talk of rights and duties is a problematic way of discussing the ethical issues.

Second, members of the current generation need not maximize the income of their descendants, with or without a discount factor. They are not even required to ensure future income levels equal to their own. We would not necessarily consider it irresponsible for the extremely rich to leave their children only moderately rich. So members of the current generation are not truly trustees who are morally obligated to preserve the entire trust for future generations.[42] Responsible individuals do at-

41. See Epstein, *supra* note 33, at 1472–1477, 1483 (suggesting use of interfamilial obligations as a baseline).

42. But cf. Edith Brown Weiss, "The Planetary Trust: Conservation and Intergenerational Equity," 11 *Ecology L.Q.* 495 (1984) (arguing that the current generation holds the natural resources of the planet in trust for future generations and must act as prudent "trustees" for future beneficiaries, taking care to preserve the "corpus").

tempt, however, to ensure that their descendants can enjoy a decent standard of living, at least if this can be done without extreme self-sacrifice. You would have grounds for complaint if your great-grandparents had taken actions that consigned you to poverty so they could live a life of luxury. You might also be properly bitter if they carelessly smashed irreplaceable family heirlooms. Again, it might not be proper to say that they had violated your "rights," but they clearly would have acted irresponsibly.

Third, members of the current generation clearly are felt to have a more compelling obligation toward the next generation (and perhaps at least to young grandchildren) than to succeeding generations. This observation is borne out by some empirical studies.[43] Members of the current generation would be criticized if they did not give the long-term welfare of their own children substantial weight. Any large discount factor would significantly undercut this responsibility to the next generation.

When we get beyond the next generation, discounting is not a particularly useful technique. Our responsibility to later generations is to ensure them a minimum level of welfare, rather than to maximize the total GDP over time. As a practical matter, it is also doubtful that we can project benefits with even minimal confidence over long periods (i.e., over a century). Even if we could predict some benefits with a degree of accuracy over such long periods, it is also doubtful that today's generation would agree to make severe sacrifices simply to create marginal improvements in the welfare of distant generations. What we can realistically attempt is to avoid substantial risks of future disaster to remote descendants. With only a few exceptions, these risks will also pose dangers to the next generation as well. Thus, these very long-term effects will usually be subsumed under our concern for the next generation.

As to the next generation or so, we might reasonably apply some discount factor. Any discount factor should not be too high, since it must accord significant weight to the interests of the next generation. In particular, the discount rate should not significantly exceed the expected long-term rate of economic

43. See, e.g., Cropper et al., *supra* note 19, at 1412–1413; Cropper and Portney, *supra* note 18, at 375.

growth; otherwise, even the expected elimination of most future GDP would be discounted to a low present value over periods of only decades.[44] Even if it would occur late in our children's or grandchildren's lives, we do not want to ignore a potential global catastrophe. Practically, this requires a discount rate no greater than 1 or 2 percent.

We can't view our own society objectively when we think about institutional design. It's helpful to imagine someone facing the problem of designing institutions to govern a closed society over an extensive period. One of the standard science fiction plots involves multigeneration interstellar travel. At the kinds of speeds that we might reasonably hope to obtain, it could take a century or more to reach even nearby stars. One solution would be to create mini-societies, with multiple generations living and dying on board during the journey. Imagine you were in charge of designing such a society. As an outsider, you have no reason to favor the current generation over later generations. (Actually, if there is any particular generation you might want to favor, it would be the one alive when the ship arrives, far in the future.) So you might want to establish social institutions that ensured strict equality between generations. On the other hand, you could hardly expect the people on shipboard to be equally impartial. Quite likely, they would have a tendency to favor their own generation or their children, instead of worrying about the far future. Indeed, unlike you, they may view their own lives as centrally important (whereas you view them as links in the chain). From your external point of view, all generations are equal, but from the perspective of the ship's passengers, no gen-

44. As Professor Lind explains:
Suppose we select a social discount rate based on present consumer rates or the rate of return on investment. Historically, any such rate is likely to exceed the rate of growth of the economy, often by a large amount. Then the basic arithmetic of exponential growth applied in a cost-benefit analysis implies that, regardless of how small the cost today of preventing an environmental catastrophe that will eventually wipe out the entire economy, it would not be worth this cost to the present generation if the benefits in the future are sufficiently distant. To most of us, this would seem a highly questionable if not immoral public policy.

Lind, "Reassessing," *supra* note 14, at S-20.

eration can be expected to take such a detached view. In designing the ship's institutions and infrastructure, this difference in perspective is something you would need to take into account. You don't want the residents to use up all the ship's resources somewhere along the way, and you can certainly try to instill a responsible attitude toward the future. Still, it would be foolish to expect complete altruism and absolute self-control.

This thought experiment also brings us once again to the issue of regulatory sustainability. It may be possible to get some people, some of the time, to make extraordinary sacrifices for future generations. We cannot, however, expect society as a whole to sacrifice present welfare indefinitely. Yet temporary sacrifices to save the environment will do little good if they are undone by later generations. So the principles governing environmental responsibilities to future generations will be unavailing unless they are fashioned in a way that ordinary people can be expected to find acceptable over time. Durable environmental protection requires a realistic appreciation of the extent of the sacrifices that we can reasonably expect to be made for later generations.

The Planet as a Long-Term Investment

So far, we have been concerned about the problem of discounting benefits that will be experienced by future generations. This is a separate problem from evaluating opportunity costs. Lawrence Summers has invoked opportunity costs as an argument for using a high discount rate for benefits accruing to future generations. He argues that standard "public non-environmental investments like sewage-treatment facilities, education programs, or World Bank transport projects have returns of more than 10%," while private investors require returns of 15 percent or more on projects. He concludes that these other options provide the benchmark for assessing environmental projects:

> Once costs and benefits are properly measured, it cannot be in posterity's interest for us to undertake investments that yield less than the best return. At the long-term horizons that figure in the environmental de-

bate, this really matters. A dollar invested at 10% will
be worth six times as much a century from now as a
dollar invested at 8%. . . .[45]

In using the higher discount rates to measure opportunity
cost, Summers is implicitly assuming that the actual alternative
investments are projects having benefits that compound over a
century. (In effect, he is assuming that environmental investments are "completely crowding out" the other public and private investments over the course of a century.) This is unrealistic.
If the nonenvironmental project has a twenty-year life (and the
typical private project is probably much shorter), then at the end
of the twenty years, the public may receive a high return on the
investment. Often, however, only a small part of that return will
be voluntarily reinvested in a new project.[46] Using the more appropriate "shadow price of capital" approach, the proper discount rate to be used over the century is probably much lower
than the return on any single short-term project.

Although Summers does not directly address this point, he
may have in mind a different scenario. The higher discount rate
would be appropriate if we could make a binding commitment
today to invest in higher return projects and to reinvest all of the
proceeds of the projects in similar new projects. The problem
is that we cannot make meaningful irrevocable commitments
regarding government (let alone private) actions over many
decades.

For precisely this reason, environmental investments may offer
a useful opportunity for precommitment. We may obtain higher
returns for the next generation by making investments today that
pay lower annual returns, but over longer periods. In this re-

45. Lawrence H. Summers, "Summers on Sustainable Growth," *Economist*,
May 30, 1992, at 65, 65. OMB has subscribed to the same rationale. See Ran-
dolph M. Lyon, "Federal Discount Rate Policy, the Shadow Price of Capital and
Challenges for Reforms," 18 *J. Envtl. Econ. & Mgmt.* S-29, S-32 (1990).

46. For example, Lind uses 0.2 as the marginal propensity to save. See Lind,
"The Rate of Discount," *supra* note 22, at 447. The underlying insight is that
individuals are likely to consume at least some of their (real) investment income,
rather than reinvesting all of their returns. Indeed, even if they "roll over" their
investments, they may increase their consumption from other sources of income
because of their greater wealth.

spect, social decision making may properly incorporate some of the procedures used by individuals who make different investments at different interest rates in the interest of precommitment. Environmental protection may be the societal equivalent of the "Christmas club," in which we invest at low returns simply to protect ourselves from wavering commitments (here operating as a whole society, rather than as individuals). Eliminating carcinogens may be a psychologically appealing savings plans.

It may be easier to protect a rain forest or the ozone layer— which might produce a 2 percent annual return over a century at the cost of $1 billion in current consumption—than to give $1 billion to the World Bank now *and* commit both ourselves and our descendants to fully reinvesting all of the benefits of bank projects. The rain forest would be more easily preserved intact than a government fund because of its vividness as a tangible symbol of the heritage of "capital" passed down between generations. Similarly, a "sustainability" requirement, which would require maintaining the world's "stock of natural capital,"[47] may be justified as a method of maintaining intergenerational savings.

Summers is correct that over a short period of time other kinds of investments may be more productive than environmental preservation. (I think he is also right to the extent that he is implying that environmental issues and economic development are connected; this is implicit in the concept of "sustainable" regulation.) The fact that other investments currently provide higher returns is not necessarily a decisive consideration, however. The "environmental tortoise" may gradually overtake the "economic hare" if it is easier to sustain multigenerational commitments to environmental preservation than to other kinds of investments.

There is a more general point here. In considering opportunity costs, we must consider only other opportunities that might actually be implemented. We should choose among the most desirable of the *feasible* alternatives. In the interest of environmen-

---

47. See Pearce and Turner, *supra* note 11, at 225. For a more general discussion of sustainability as an ethical stance, see Eric Freyfogle, "Should We Green the Bill?" 1992 *U. Ill. L. Rev.* 159, 162–163.

tal protection, people might be willing to give up $1 billion of current consumption. This does not necessarily mean that they would be willing instead either to pay an extra $1 billion in taxes to finance World Bank development projects or to save an extra $1 billion for private investment. Instead, absent the environmental regulation, they might simply consume the extra $1 billion. Thus, in considering the opportunity cost of environmental decisions, we must be realistic in determining which alternatives are politically and socially feasible.

To return to the family context we explored earlier, parents may wish to ensure their offspring's inheritance, but they may also find it difficult to put aside savings for this purpose. They may find it easier (though in some sense less efficient) to hold onto family heirlooms, even if those heirlooms appreciate more slowly than some other investments. A stewardship ethic may function as a way of committing our own generation to saving for future generations when we find it difficult otherwise to carry out long-term plans that we find ethically desirable.

Such a stewardship ethic does not require giving heavy weight to the interests of far distant generations; it is enough if we are determined to maintain our global inheritance largely intact during our children's lives, leaving it to them to apply the same ethic to their own successors. If we fulfill our responsibility to the next generation, including the teaching of an environmental ethic, more distant generations will be provided for without the necessity of an unrealistic emphasis on the far future as we make our own decisions. Like runners in a relay race, we may do best when we concentrate on passing the baton to the next runner, leaving the rest of the race to the succeeding runners.

Some readers may think that my approach is shortsighted because it stresses commitments to nearby generations over those further into the future. I do not believe that this approach slights the long-term interests of the human race. We are talking about planning for the full life spans of our children. This is a substantial increase over the time horizons typically used by today's politicians and businesses.

Moreover, I have doubts about the workability of any planning horizon much longer than the life of the next generation. It would be very difficult to motivate individuals to sacrifice for returns that are delayed far beyond the life spans of their own children. It would be even more difficult to design democratic institutions that would keep a given social program in place for such long time periods. So as a practical matter, any policy choice made today has only a finite period of effectiveness. Finally, even if it were possible to "lock in" policy choices for many generations, it is doubtful that this would be desirable. Our information about long-term policy impacts is unreliable. As discussed in the next chapter, we have every reason to think that later corrections will be in order. It would be a mistake for us to try to forecast and solve all the problems of our grandchildren and subsequent generations. We will do well enough if we leave them a livable world and well-designed institutions with which to make their own choices.

I earlier rejected the idea that the current generation truly is a trustee for the overall welfare of future generations. Nevertheless, this analysis suggests that it may be useful for society to think in terms of a more limited stewardship. First, the current generation may have difficulty meeting its own savings goals for future generations. It may well be useful to treat aspects of the ecosystem as if they were family heirlooms, as a technique of increasing savings. Second, the current generation has at least a responsibility to leave later generations the minimum requirements for decent lives, which means avoiding any severe, irreparable environmental damage. Depending on the level of sensitivity of the global ecosystem, this may substantially constrain current decisions.

In this chapter, I have more or less taken it for granted that we should give serious attention to the needs of our future selves, even several decades in the future, and to the needs of our descendants. Of course, we have an instinctive desire to protect the welfare of our current selves and of our children. In some sense, the question of the validity of these desires does not even arise; they are simply part of what makes us human. But our instincts are not designed for planning over long time spans, and we are not psychologically compelled in the same way to take the future into account when it extends so far beyond our immediate experiences. Why not leave the future, especially the far future, to take care of itself? One answer may be our desire to transcend the present by situating ourselves in a community that stretches from the past, those who created the society in which we were born, to the future, those to whom we will bequeath the fruits of our own efforts. Thus, the reasons to take the future seriously are much like those for taking the past seriously. We will always remain rooted in the present, but at least we can orient ourselves in a broader span of history.

For shorter periods of time, discounting can be used to account for future environmental benefits. Because our society seems to be too present-oriented in some regards, we ought to use a relatively low discount rate so that the government can take a longer view. Technically, we can make intertemporal comparisons by using the "shadow price of capital" method. For those who prefer not to discount environmental and health benefits, we can use similar methods to project the costs forward into the same time period as the benefits. For longer time periods, we can use the idea of stewardship as a shorthand for our responsibilities to future generations.

The problem discussed in this chapter—time lags of decades or centuries—may seem difficult enough. But until this point, I have been putting aside one of the most difficult complications. In discussing discounting, as well as in the earlier discussion of risk management, I have mostly ignored one of the overarching problems of environmental law. The discussion has proceeded on the assumption that we have a good grasp of the risk posed by various environmental problems. Unfortunately, this assumption is resoundingly false, as is especially obvious when we think about effects stretching over centuries. Environmental law involves a series of problems whose dimensions are only poorly understood—and of course, very long-range effects involving multiple generations are even less well understood. The problem of uncertainty is the focus of the next chapter.

CHAPTER SIX

Dynamic Environmental Regulation

Reserve Mining raised several difficult problems. We have already examined two of those problems: how to make trade-offs between risk and cost and how to take into account the long time lag before control measures could affect cancer rates. But the overriding problem faced by the court was uncertainty about the very existence of the risk. After carefully reviewing all the evidence, the best the Eighth Circuit could conclude was that the hazard "can be measured in only the most general terms as a concern for the public health resting upon a reasonable medical theory."[1]

Even today, risk assessment is shrouded in uncertainty. Justice Stephen Breyer, for example, reports that "[t]wo scientifically plausible models for the risk associated with aflatoxin in peanuts or grain may show risk levels differing by a factor of 40,000."[2] There is uncertainty connected with each step in estimating risks. If there are ten independent steps and each is uncertain by a factor of two, the total risk is uncertain by a thousand.[3] Such uncertainty is not uncommon in environmental law and presents a formidable challenge to decision makers.

---

1. *Reserve Mining v. United States,* 514 F.2d 492, 536 (8th Cir. 1975).
2. Stephen Breyer, *Breaking the Vicious Circle: Toward Effective Risk Regulation* 45 (1993).
3. Id. at 110 n.70.

In thinking about uncertainty, it's also important not to lose sight of the time dimension. At any given time, our knowledge may be sharply limited, causing severe difficulties in addressing environmental issues. But the frontiers of scientific ignorance shrink over time. We need strategies that exploit the possibility of obtaining better information in the future.

Some of those strategies are discussed in this chapter. In the short run, we are sometimes forced to make decisions with information too crude to allow even rough estimates of risk. Prudence may dictate that we regulate nonetheless, but it may also dictate that the response be tempered, rather than draconian. These situations of maximum uncertainty are perhaps the most difficult for a decision maker. But we should bear in mind that our state of uncertainty is not fixed for all time. Various methods exist to obtain new information and use it as effectively as possible. In particular, we need to be careful about making irrevocable decisions if new information may shed later light on the situation. Finally, given our state of uncertainty, we are bound to overreact on occasion, and we need a mechanism for regulatory retrenchment when we learn that our first response was excessive.

My goals in this chapter are limited. An analogy to pollution regulation may clarify the nature of the project. One of the conventional regulatory technologies is BAT (best available technology). Rather than inventing new technologies, a task well beyond the scope of its expertise, the EPA sets standards based on the best technology currently used in the industry or allied fields, or sometimes even on technology still in the demonstration stage. BAT is not meant to be revolutionary, but it does sometimes require major changes for many firms. Similarly, my purpose in this chapter is not to create entirely new methods of regulation, but to identify the best methods currently in use in environmental law or related fields, including some that are still in the "demonstration stage." Thus, in the spirit of BAT and other pollution control methods, one might say that the goal of this chapter is to identify BART (best available regulatory techniques). Just as there is often debate about the EPA's choice of BAT, my choice of BART will undoubtedly be subject to some dispute. To complete the analogy, the chapter might be thought of as merely the opening salvo in the "notice and comment" process, which will surely result in improvements.

Thus, the proposals made in this chapter are intended as starting points for discussion, not as complete action plans. Designing approaches to environmental protection is a complex matter, involving difficult judgments about the manifold effects of any proposal. Questions such as the allocation of authority between courts and agencies, or between the federal and state governments, have broad systemic implications that cannot be fully explored here. Even if the basic directions of the proposals are found to be valid, they would require much detailed development in order to be implemented. My purpose, then, is not to dispose of the hard questions of regulatory design in the course of a single brief chapter. Rather, it is to illuminate how we might creatively respond to the pervasive scientific uncertainty within environmental law.

Obviously, there are many competing considerations in designing a regulatory system. Although I do not argue that the considerations discussed in this chapter are decisive, they do deserve heavy weight in the analysis. Administrative law has tended to focus on the virtues of a given approach at a specific moment in time. Thus, the analysis is static, rather than dynamic. My focus in this chapter is on the dynamic nature of decision making, particularly on the need to foster the ability of the regulatory system to engage in learning over time. If, as I have argued in the last chapter, we need to take a long-term view of environmental protection, dynamic factors may ultimately outweigh static ones, just as a high enough rate of return on one project must in the long run outweigh the lower cost of another. A system in which we learn from our mistakes may ultimately outperform one that makes more careful decisions in the short run, but is unable to improve in the light of experience. Thus, while the need for dynamic learning is not the only factor to be considered in designing regulatory mechanisms, it should play a major role in the analysis.

Living with Radical Uncertainty

It is humbling to contemplate the extent of our ignorance. The width of the statistical "confidence intervals" in risk estimates is sometimes startling. For instance, one risk analyst reported that a pesticide would cause twenty-seven cases of cancer over a seventy-year period; the statistical "confidence interval" turned out to range from zero (no cancers) to sixty thousand. As Tom McGarity observes, "This vast interval spoke volumes about the confidence with which the analyst made his original prediction of twenty-seven cancers."[4]

Uncertainties in risk estimates are sometimes due to theoretical disputes such as how to model the dose-response curve. Other uncertainties stem from the difficulty of interpreting limited data. For example, calculations of the lifetime risk of benzene to workers ranged from one to twenty-five per thousand. The disagreement turned on different estimates of the background rate for a particular form of cancer in Turkey, where one of the key studies had been done.[5]

Some experts question whether the elaborate procedures used to calculate risks are really meaningful. They argue that "risk assessment has become too formalized and mechanical in light of the limited data" and that "[l]ittle is gained from the sophisticated massaging of weak data."[6] One expert compares risk assessment to the method used to weigh hogs in Texas: "Down there, they put the hog in one pan of a large set of scales, put rocks in the other pan, one by one, until they exactly balance the weight of the hog. Having done that very carefully, they guess how much the rocks weigh."[7]

Perhaps McGarity exaggerates when he describes risk disputes as battles where "numbers that nobody takes very seriously are tossed around like so many hand grenades to defend positions staked out for other reasons."[8] Still, the seemingly precise estimates of lost statistical lives are more speculative than they may seem. No matter what methodology we use to make decisions, we must live with this uncertainty until scientific progress allows more reliable estimates.

Similarly, estimating the likelihood that a species will become

4. Thomas McGarity, *Reinventing Rationality: The Role of Regulatory Analysis in the Federal Bureaucracy* 135 (1991).

5. John Graham et al., *In Search of Safety: Chemicals and Cancer Risk* 164–165 (1988). Other technical matters, like how to classify various kinds of tumors, are often also hotly disputed.

6. Id. at 177.

7. Breyer, *supra* note 2, at 108 n.58.

8. McGarity, *supra* note 4, at 50.

extinct may be extremely difficult. We often lack basic information about the species such as knowledge of its feeding and reproductive habits, its susceptibility to diseases, and even its current population. Mathematical models of the extinction process are still crude and are likely to underestimate risks. Even after we decide that a species is endangered, we may not know enough to mount a successful response. Often, we may not be sure that the species is threatened until it is too late.[9] Indeed, even with well-studied stocks of fish, we seem unable to determine safe levels of fishing.[10]

Critics of cost-benefit analysis often worry that quantitative methods will "dwarf soft variables"—that is, that intangible values, which can't be easily quantified, will be given short shrift, as opposed to hard variables with definite numbers. As it turns out, however, *all* the variables are soft.

Even the cost of compliance, usually taken as a straightforward economic measurement, is subject to great uncertainty. Data about out-of-pocket costs is often unreliable, with an uncertainty of plus or minus 30 percent (that is, from 70 to 130 percent of the estimate, almost a factor of two). Sometimes costs have been badly overestimated, occasionally to an amazing extent. McGarity reports that "a retrospective look at the costs of complying with OSHA's vinyl chloride standard found the actual costs were only about 7 percent of predicted costs."[11] Even if we did have accurate predictions of direct compliance costs, they would provide a poor measure of the overall cost to society. Overall social cost depends on how the economy adjusts indirectly to the regulation through various price changes, productivity effects, and so forth. One effort to pinpoint some of these effects found that at different times the true economic cost of environmental regulation ranged from 50 to 150 percent of the apparent compliance cost.[12]

9. See National Research Council, *Science and the Endangered Species Act* 170–185 (1995).

10. See Donald Ludwig et al., "Uncertainty, Resource Exploitation, and Conservation: Lessons from History," 260 *Science* 17 (1993).

11. McGarity, *supra* note 4, at 127, 137.

12. See Raymond Kopp et al., *Discussion Paper: Cost-Benefit Analysis and Regulatory Reform: An Assessment of the Science and the Art* 25–26 (Resources for the Future Jan. 1997).

Given these uncertainties, it is hard to say whether current environmental regulations are cost-justified. The estimates of total costs for a single year (1988) range from $55 to $77 billion, while benefit estimates range from $16 to $135 billion.[13] This means that for every dollar we invested in environment protection, we got back somewhere between $0.21 and $2.27. We were either losing 80 percent of our investment or more than doubling our money, we don't know which! And in particular, we have no idea whether the total cost-benefit analysis for environmental protection is favorable or unfavorable.

To use a phrase coined by Winston Churchill, it is sometimes hard to avoid the feeling that environmental protection is a "riddle wrapped in a mystery inside an enigma."[14] How can we possibly make sound public policy under these circumstances?

Although the uncertainty surrounding environmental policies is humbling, we are not completely in the dark. We do know that the likely level of risk is considerably larger in some situations than in others. We have at least crude information about compliance costs and rough notions about what kinds of financial sacrifices people are willing to make for safety. We need to find ways to use what information we have to make sensible decisions, without forgetting the uncertainty surrounding those decisions.

The softness of our information base reinforces the argument made in chapter 4 for placing heavier reliance on feasibility analysis than on cost-benefit analysis. Cost-benefit analysis, because it is more quantitative and formalized, puts higher information demands on the analyst. We have to know not only that the risks are significant, but also just how high they are; not only that the costs are feasible, but also just what they will run; not just that a number of deaths over a given period of time is too high, but also what their monetary value is and how to discount it to present value. And we need to know *all* these things accurately enough to compare the magnitudes of the relevant figures. It will some-

---

13. Robert W. Hahn and John A. Hird, "The Costs and Benefits of Regulation: Review and Synthesis," 8 *Yale J. on Reg.* 233, 256 (1990). For a more optimistic EPA estimate, see Alec Zacaroli, "General Policy: Final Report Finds Benefits of Air Act at Least 10," 28 *Env't Rep.* (BNA) 1243 (1997).

14. *Bartlett's Familiar Quotations* 620 (Justin Kaplan ed., 14th ed. 1992).

times be clear that a cost-benefit analysis comes out one way or the other, but more often, we will simply be left unsure at the end of the day. Someday, with a better information base, we may obtain more precision in cost-benefit analysis. But we are a long way from reaching that point today. For now, we can best use cost-benefit analysis as an auxiliary check on the reasonableness of our feasibility analysis.

Given such soft data, we have little recourse but to rely heavily on the EPA's expertise and experience to assess risks. We can take various measures such as encouraging use of peer review to strengthen the agency's ability engage in risk assessment. Still, there may always be something of a "black art" to risk assessment, just as there is to other difficult professional judgments. To quote the authors of a particularly careful case study of risk assessment, the "frontiers of science are often ambiguous, and disagreements are not readily resolved by objective or even consensual means. Instead, intuition, craftsmanship, and judgment are critical."[15]

We should not be overly confident of the ultimate validity of these expert judgments. Such judgments may sometimes come perilously close to guesswork, and experts have biases of their own. But what is the alternative? About the only thing we can say is that the opinions of experts about hard technical issues are more likely to be correct than are the judgments made by politicians or lay people. In short, heavy reliance on experts is an uncomfortable strategy, but probably unavoidable.

Although the problem of scientific uncertainty is daunting, we should not allow ourselves to be overwhelmed by it. Risk estimates may not be completely reliable, but at least they can often give us a rough indication of the possible seriousness of a situation. Even if they lack ultimate validity, quantitative techniques can provide an appreciation of the dimensions of a problem and of the kinds of responses that would be reasonable. When these techniques indicate the existence of a significant risk, even if we cannot determine the exact magnitude of the risk with certainty, the hybrid approach of chapter 4 calls for reducing the risk to the extent feasible (subject to the gross disproportionality test).

15. Graham et al., *supra* note 5, at 187.

The most difficult situation is one in which the agency itself does not feel capable of making meaningful quantitative judgments, even in a very rough way. Even with feasibility analysis, risks or compliance costs may be so uncertain that we are unsure whether to regulate. The European Union (EU) has adopted the "precautionary principle" in response to this problem. This principle is variously explained as being based on risk aversion or on skepticism about the capacity of the environment to assimilate pollution.[16] The precautionary principle is now part of the Maastricht Treaty, which is the current constitutional framework of the EU. It also appears in the Rio Declaration on international environmental law.[17] In part, the precautionary principle is based on experience with unpleasant environmental surprises. As one observer explains, "New evidence of the irreversible harm done by compounds whose pollutive impact had previously been thought insignificant," such as chlorofluorocarbons, promoted the view that "it was now necessary to take action in advance of conclusive scientific proof that would connect observed degradation with particular substances or sources of pollution."[18]

The precautionary principle is at least a useful reminder that we should be cautious in gambling with human lives or ecological integrity. In this sense, it is less a method of decision than a form of coaching, like reminding a teenager to drive carefully. At this level, the principle is helpful, but not terribly illuminating. American law provides some analogs to the precautionary principle, however, that may have somewhat more specific content.

One possible, but ultimately unappealing, analog is the worst-case analysis that was formerly required in environmental impact statements. When the environmental risks of a project were uncertain, the agency had to include a description of the worst case in the impact statement. This requirement had a checkered career. It was once mandated by executive order, but after the order was withdrawn, some courts required agencies to engage in

16. Richard Revesz, "Notes and Questions," in *Foundations of Environmental Law and Policy* 331 n.10 (Richard Revesz ed. 1997).

17. See id. (citing Rio Declaration, Principle 15); Onno Brouwer et al., *Environment and Europe: European Union Environment Law and Policy and Its Impact on Industry* 9, 34–36 (1994).

18. Brouwer et al., *supra* note 17, at 35.

a worst-case analysis anyway until the Supreme Court held that no such analysis was required by the statute.[19] Whatever may be said of the worst-case analysis as a disclosure requirement, it makes a very poor standard for regulation. Basing regulation on the worst case would be taking Murphy's Law ("whatever can go wrong will go wrong") to a paranoid extreme.

This approach would result in a colossal waste of resources when the feared disaster evaporates. Ironically, society's actual risk could be increased to the extent that unregulated alternatives turn out to be more dangerous than the regulated risk.[20] In the long run, such an approach would probably be damaging to environmentalism. Like the boy who cried, "Wolf," the government would continually be announcing an impending catastrophe that it would later have to admit was a false alarm. How long would people continue to take environmental problems seriously after a few rounds of this cycle?

A more useful interpretation of the precautionary principle is that it shifts the burden of proof to the polluter, which must show that its activities are harmless.[21] This interpretation bears a strong resemblance to doctrines developed by the D.C. Circuit Court in construing the Federal Insecticide, Fungicide, and Ro-

---

19. *Robertson v. Methow Valley Citizens Council,* 490 U.S. 332, 354 (1989). The case, which involved the possible impact of a planned ski resort on a herd of mule deer, was decided in large part in deference to the position taken by the executive branch. Instead of the worst-case requirement, the Court upheld a replacement regulation. The new regulation directed agencies, in the face of "unavailable information concerning a reasonably foreseeable significant environmental consequence," to prepare "a summary of existing credible scientific evidence" and to present an "evaluation of such impacts based upon theoretical approaches or research methods generally accepted in the scientific community." See id. (quoting regulation).

20. See generally Frank B. Cross, "Paradoxical Perils of the Precautionary Principle," 58 *Wash. & Lee L. Rev.* 851, 862–882 (1996).

21. Philippe Sands, "The 'Greening' of International Law: Emerging Principles and Rules," 1 *Ind. J. Global Legal Stud.* 293, 297–302 (1994), reprinted in *International Environmental Law Anthology* 21, 21–22 (Anthony D'Amato and Kirsten Engel eds., 1996). See also Thomas McGarity and Sidney Shapiro, "OSHA's Critics and Regulatory Reform," 31 *Wake Forest L. Rev.* 587, 630 (1996) ("Why . . . should the presumption be that nothing can be done to protect workers, unless the government can meet the affirmative burden of showing that the benefits . . . outweigh the costs. . . ?")

denticide Act, the federal law requiring registration of pesticides with the EPA. If doubts about safety arise later, the registration can be first suspended and then canceled if the doubts are confirmed. As a result of a long-running dispute between the EPA and the Environmental Defense Fund, a series of opinions by Judge Harold Leventhal announced rules regarding the burden of proof in this process.

A brief summary will give the flavor of these rules. Once the EPA issues a notice that it will seek cancellation of a pesticide, a presumption arises in favor of suspending use of the pesticide during the proceeding. When the EPA does suspend use of a pesticide, another presumption arises in favor of permanent cancellation if no benefit from using the pesticide (as opposed to alternatives) is shown or if animal tests show that the chemical causes cancer. Moreover, if evidence shows that the one mode of exposure is hazardous, a presumption arises that all modes are hazardous until proven otherwise. Thus, for example, if inhalation of a chemical is dangerous, there is a presumption that ingestion of the chemical is also dangerous.[22] In *Reserve Mining*, this would have meant a presumption that asbestos is a hazard in drinking water, since inhaling it is extremely dangerous.

Adjusting the burden of proof may also be appropriate in a more technical setting. The statistical tests used by scientists are mostly designed to exclude a particular kind of error. They minimize the likelihood of mistakenly concluding that an effect is real when it actually does not exist (Type I error). The higher we set the threshold for recognizing that something exists, however, the more likely we are to miss it when it really is present (Type II error). If you're extremely skeptical, you will rarely be persuaded by a false argument, but you'll dismiss a lot of true, but debatable, information. (For example, if we use the standard .05 percent level of statistical significance, we only make the first kind of error one time out of twenty, but we may make the second

22. See *Environmental Defense Fund, Inc. v. EPA [Aldrin and Dieldrin]*, 465 F.2d 528 (D.C. Cir. 1972); *EDF v. EPA [Aldrin and Dieldrin II]*, 510 F.2d 1292 (D.C. Cir. 1975); *EDF v. EPA [Heptachlor and Chlordane]*, 548 F.2d 998 (D.C. Cir. 1976). Since the names of the parties are identical, these cases are conventionally distinguished from each other by including the name of the pesticide in the citation.

error more frequently, failing to observe an effect as much as 90 percent of the time it is present.) Thus, according to the National Research Council, we may want the experts to adjust their statistical tests when the second kind of error is especially serious, like the failure to detect that a species is endangered. This does not necessarily mean reversing the burden of proof, but it does mean adjusting the threshold at which we are willing to take action.[23]

Although the idea of burden-shifting is promising, it does have problems. We must not overuse this technique or set the burden of proof too high. Given the high degree of uncertainty, there are limits to how effectively polluters can justify their activities. Setting the burden too high has the same effect as adopting the worst-case scenario approach: once concerns are raised, we will always in effect be assuming the worst. Regulatory agencies should normally have the burden of justifying restrictions on private parties. If such an analysis is possible, the agency should be required to show that the risk is significant and that the remedy is feasible. This should be done as quantitatively as possible, given the state of the art. But burden-shifting can function usefully as a kind of tiebreaker. The agency's hands should not be tied when it has evaluated all of the available evidence, but cannot make a confident risk estimate. Instead, it should be free, in its best judgment, to implement reasonable regulations as a precaution against environmental hazards.

Courts have varied in their willingness to tolerate such qualitative judgments about risk. In *Corrosion Proof Fittings,* the court took the position that neither it nor the EPA could make a sensible decision without quantifying relevant variables. The *Reserve Mining* court took a more prudent approach, which seems more consistent with our society's environmental baseline. If people have a presumptive right to a safe environment, then we cannot rest easy in tolerating potentially dangerous forms of pollution, even if we cannot yet confidently quantify the risks. Given the information before it, I believe the *Reserve Mining* court was right to conclude that the "risk to public health is of sufficient gravity to be legally cognizable and calls for an abatement order on rea-

23. National Research Council, *supra* note 9, at 162–170.

sonable terms."[24] It would have been better to have a fuller knowledge of the risk and the possible remedies, but decisions must sometimes be made without this luxury.

The Moving Knowledge Frontier

Although our information about environmental risks is still limited, we are faced with rapidly changing knowledge regarding environmental hazards. A decision that seems entirely correct today may turn out to be completely wrongheaded tomorrow.

Again, *Reserve Mining* provides an apt illustration. Given a choice between requiring land disposal and doing nothing, the *Reserve Mining* court was probably correct to opt for land disposal. But before land disposal was actually implemented, it may have become unnecessary. By 1977, a water filtration system had been installed in Duluth that removed 99.9 percent of the asbestos fibers.[25] This meant a reduction in the risk by a factor of one thousand, down to something like one death every six hundred years (1.5 deaths per year divided by a thousand). Whatever the exact definition of "significant risk" may be, this risk clearly did not qualify. For purposes of comparison, this risk was hundreds of times smaller than the expected number of deaths each year from high school football injuries, aircraft accidents, or drownings in Duluth.[26]

Clearly, it wasn't worth spending $200 million or more to eliminate such an insignificant risk. In a society that long tolerated over a hundred deaths per year from smoking in a city the size of Duluth,[27] requiring Reserve to spend a fortune to eliminate such a minuscule risk seems hypocritical. Under a cost-benefit analysis, land disposal would be justified only if we used

24. *Reserve Mining v. United States,* 514 F.2d 492, 500 (8th Cir. 1975).
25. Eunice E. Sigurdson, "Observations of Cancer Incidence Surveillance in Duluth, Minnesota," 53 *Envtl. Health Persp.* 61, 62 (1983). The plant cost $6.9 million, of which Reserve ultimately paid $1.1 million. UPI, Regional News, Oct. 21, 1981.
26. B. T. Mossman et al., "Asbestos: Scientific Developments and Implications for Public Policy," 247 *Science* 294, 299 (1990).
27. This is also derived from the Mossman article, *supra* note 26, dividing by ten to convert from rates per million to rate per hundred thousand (the size of Duluth).

a zero discount rate and also put a valuation of around $200 million on each life saved. This is far out of line with the costs imposed by most government regulations per life saved, not to mention being vastly in excess of what people demand to compensate them for occupational risks, even when they are represented by unions.[28] It seems particularly bizarre to spend so much to remove the last microscopic traces of asbestos from the drinking water when such common products as aspirin and rice may contain much more asbestos.[29]

Perhaps Reserve was causing sufficient ecological damage to the lake to justify conversion to land disposal. That was the initial issue in the case, which never went to trial because of what appeared to be a more urgent concern about public health. More recent scientific evidence raises serious concerns about disruption of aquatic ecosystems and significant harm to lake trout and other fish.[30] The sheer symbolic outrage of dumping 67,000 tons of waste a day in a pristine lake may well be enough to justify a switch to land disposal—if not immediately, at least after a reasonable time. But in retrospect, if land disposal is to be justified, the justification should not rest too heavily on concern about drinking water, for that concern may have been fully addressed by water filtration.

Of course, this was far from being the only unexpected development in the *Reserve Mining* case. Perhaps the most notable was the sudden emergence of the public health issue on the eve of

28. Even if the filters removed only 90 percent of the asbestos, the risk levels would drop to the point where it would be hard to maintain that there was a serious health concern. Professor Nicholson suggests that "standard flocculation and sedimentation techniques can reduce asbestos concentration by about 90%" at "relatively modest cost." William J. Nicholson, "Human Cancer Risk from Ingested Asbestos: A Problem of Uncertainty," 53 *Envtl. Health Persp.* 111, 112 (1983). If risks can be reduced by that amount, some other factor in the cost-benefit analysis would have to be increased by a factor of ten to compensate. Similarly, whether the remaining risk is "significant" seems at least debatable.

29. According to MacRae, "[A]n ordinary diet including one beer per day, some rice and three aspirin" would be the equivalent of drinking three liters of water per day having about 2,500 million fibers per liter. K. D. MacRae, "Asbestos in Drinking Water and Cancer," 22 *J. Royal C. Physicians London* 7 (1988). That is about twenty-five times the peak level in Duluth drinking water.

30. James Webber and James Covey, "Asbestos in Water," 21 *Critical Reviews in Envtl. Controls* 331 (1991).

trial, due to the new scientific understanding that asbestos was a carcinogen. Because of scientific advances, major changes in our understanding of environmental problems have been frequent. We have no reason to expect that our current understanding of environmental risks will be immune from revision.

Sometimes new knowledge demotes risks previously considered more serious. In the early 1970s, toxic chemicals were considered a major cancer threat, and leaking hazardous waste sites were considered a "clear and present" danger to public health. Rigorous regulation followed, in the form of strict federal controls on future waste disposal, with Superfund to clean up existing waste sites. By the time these statutes were in place, however, their scientific basis was already eroding.[31]

Dioxin exemplifies the changing scientific view of carcinogens. Dioxin was once considered the mostly deadly carcinogen in existence, even in microscopic doses. It may still retain that reputation with the public, but at this point, the scientific evidence is less clear. By 1991, based on evidence that dioxin can cause harm only after binding to certain cellular receptors, scientists argued that it might well be safe below certain levels.[32] Recent empirical data also raises questions about the dangerousness of dioxin. A factory explosion in 1976 exposed thirty-seven thousand people to high levels of dioxin. A recent study showed worrisome increases in some cancers, but the overall cancer rate was actually lower than the rate among the general population.[33] Some grounds do remain for serious concern about dioxin,[34] but the risks are cloudier than previously believed. Indeed, the most recent information suggests a similar story about ingested asbes-

31. See Brian Henderson et al., "Toward the Primary Prevention of Cancer," 254 *Science* 1131, 1137 (1991).

32. Leslie Roberts, "Dioxin Risks Revisited," 251 *Science* 624 (1991). For discussion of more recent developments, see Richard Stone, "Dioxin Receptor Knocked Out," 268 *Science* 638 (1995).

33. Keith Schneider, "2 Decades after Toxic Blast in Italy, Several Cancers Show Rise," *N.Y. Times*, Oct. 26, 1993, at B6. There is some reason to believe that dioxin actually inhibits the development of certain cancers, rendering the situation all the more confusing.

34. See Richard Stone, "Panel Slams EPA's Dioxin Analysis," 268 *Science* 1124 (1995).

tos. After the initial scare at the time of *Reserve Mining*, later evidence remains inconclusive,[35] and the World Health Organization believes no risk exists.[36] That view may be too optimistic, but it is clear in any event that the danger of ingested asbestos is less blatant than once feared.

Although dioxin and some other toxic chemicals currently seem less dangerous than once believed, scientific knowledge can also reveal new dangers. When international negotiations began in 1986, it was quite unclear whether the ozone layer was actually in any danger. Although more convincing evidence relating to the Antarctic ozone "hole" began to appear during the negotiations, it was only later that a scientific consensus emerged.[37] Even today, however, there are major uncertainties about the causal mechanisms and effects of ozone depletion.[38]

Similar uncertainties exist about other major environmental issues. For example, after a careful review of the evidence regarding the greenhouse effect, Christopher Stone found the situation much murkier than he expected:

> [H]aving recited in a draft the popular menace that the polar ice caps were ready to melt on us and so on, I waited for the authoritative backing to materialize in memos [from Stone's research assistant]. I waited in vain. The deeper into the better authorities we fished, the vaguer and more qualified the projections we landed. . . . Over the space of the few years that I have been following the research developments, all of the original, highly publicized projections of climate change

35. See John F. Gamble, "Asbestos and Colon Cancer: A Weight of the Evidence Review," 102 *Envtl. Health Persp.* 1038 (1994); Brooke Mossman and J. Bernard L. Gee, "Asbestos-Related Diseases," 320 *N. Engl. J. Med.* 1721, 1725 (1989); Richard Lemen et al., "Report on Cancer Risks Associated with the Ingestion of Asbestos," 72 *Envtl. Health Persp.* 253 (1987).

36. "U.N. Report: Asbestos Ingested in Drinking Water No Danger," *Chicago Trib.*, Feb. 25, 1994 (1994 WL 6453088).

37. See Richard Benedick, *Ozone Diplomacy: New Directions in Safeguarding the Planet* 17–18 (1991).

38. Gary Taubes, "The Ozone Backlash," 260 *Science* 1580, 1583 (1993). For a report on the current status of ozone research, see Jose Rodriguez, "Probing Stratospheric Ozone," 261 *Science* 1128 (1993).

variables have without exception crept back to much
more modest levels than in the original scare stories.[39]

Today, a broad scientific consensus exists regarding the reality of
global warming, but much remains unclear about the scope of
the problem. Although the scientific evidence is now becoming
firmer, it would be a mistake to think we already know the final
answers about global warming.[40] Similarly, despite well-founded
concerns about the loss of biodiversity, we are just beginning to
obtain basic data such as how much of the Amazon forest we are
losing.[41] Such uncertainties are no excuse for inaction, but neither should they be ignored.

It is tempting to think that *now* we finally understand environmental risks and need only to find appropriate solutions. The
reality is that we are faced with a high degree of uncertainty. But
that uncertainty is not static—scientists are constantly improving our knowledge base. This reality rather than any comfortable
illusion of scientific certainty must help shape any intelligent
strategy of environmental protection.

Change has been a constant theme in environmental law.
Looking back over the thirty-year history of environmental law,
one cannot help but be struck by just how much we have
learned. We have discovered that regulatory efforts that seemed
promising when enacted can sometimes prove disappointing in
practice. We have found that the scientific basis for environmental protection can shift quickly, as with the ozone layer. Indeed,
we have learned that our environmental agenda itself is subject
to constant revision; we seemingly learn that some problems are

39. Christopher Stone, *The Gnat Is Older than Man: Global Environment and
Human Agenda* xvi–xvii (1993) (see id. at 13–16, 20–25, for his review of the
evidence). Stone goes on to conclude, correctly in my view, that despite the substantial uncertainties, the possibility of global climate change should be taken
very seriously. See id. at 26–32. It also should be noted that longer-term predictions (over a 200- to 500-year period) remain quite gloomy. See Richard Kerr,
"No Way to Cool the Ultimate Greenhouse," 262 *Science* 648 (1993).
40. See Richard Kerr, "Greenhouse Forecasting Still Cloudy," 276 *Science*
1040 (1997).
41. See David Skole and Compton Tucker, "Tropical Deforestation and Hab-
itat Fragmentation in the Amazon: Satellite Data from 1978 to 1988," 260 *Science* 1905 (1993).

under control or are less serious than we thought, just when we also discover entirely unforeseen environmental problems.

Strangely enough, we don't seem to have adapted to the reality of this constant change. Subconsciously, we seem to assume that whereas much of what we believed five or ten years ago is outmoded, we can now make permanent decisions about environmental protection. Thus, we still seem to conceptualize environmental protection in static terms: Given the information now available, what is the best solution to a given environmental problem?

When information changes slowly, this may be the best way to think about public policy. But when the information base is itself subject to rapid change, a more dynamic approach is needed. It makes little sense to agonize over today's decision when it is likely to require revision tomorrow anyway. Moreover, given the inadequacy of our current information, developing new information is critical. Finally, because of inadequate information, predictions about a decision's effects have only limited value. Instead, we need to be more experimental, trying a lot of different things and attempting to learn from the results.[42]

In a nutshell, one of the main lessons we should learn from the last three decades is the centrality of learning to the enterprise of environmental protection. The remainder of this chapter considers ways in which environmental regulation can improve its ability to learn (raising our regulatory IQ, so to speak).

Teaching the Elephant to Waltz

Once Congress has established environmental goals, we need to design administrative systems that are responsive to additional information. One way to improve environmental learning is decentralization—moving decision making from large federal bureaucracies to the private sector or to smaller units of government. Another way is to streamline the federal regulatory process, trying to make administrative agencies less cumber-

42. For some thoughts about the relationship between complexity theory and the dynamics of environmental law, see J. B. Ruhl, "Thinking of Environmental Law as a Complex Adoptive System: How to Clean Up the Environment by Making a Mess of Environmental Law," 34 *Houston L. Rev.* 933 (1997).

some. We need to be careful not to compromise our commitment to environmental goals in the process, but intelligent reforms have the potential to provide a more responsive regulatory system.

Decentralization

Large hierarchies are not famous for responding quickly and effectively to new information.[43] We need to make regulation more nimble. Sensible pollution control requires information about the technological and economic circumstances of polluters. These conditions are subject to rapid change, and because of the inherent delays of centralized decision making, the EPA may be unable to keep up with these changes. By decentralizing environmental decision making, we may be able to obtain improved responsiveness to changing circumstances and new information.

Markets are one form of decentralization. The standard arguments for incentive schemes focus on their static efficiency. If economists are right, these schemes should do a good job of allocating responsibility for pollution control among various polluters at any given time. But dynamic efficiency may be even more important. Markets can do remarkably well at responding quickly to new information. One brokerage firm bought a supercomputer to shave two seconds off its response time for shifts in the Tokyo stock market.[44] Although this lightning response can't be considered typical of all market institutions, it does highlight how markets can force firms to learn quickly from new information. So one method to expedite learning is to organize a market.

Economists have designed ingenious incentive systems for environmental protection. One step in this direction was the system of marketable sulfur-dioxide allowances established under the 1990 Clean Air Act. Although these incentive systems are intriguing, we should not be too confident about translating that theory into practice. There are good reasons for caution. Real-

43. In the private sector, one need only consider the difficulties encountered even by a "model" corporation like IBM in a period of extremely rapid change.

44. Survey, "Frontiers of Finance," *Economist*, Oct. 9, 1993, at 4, 4 (supplement to magazine).

world implementation may raise significant enforcement problems, create barriers to entry by new firms, unduly favor some firms in the initial allocation of permits, or conflict with other goals like equity.[45] Moreover, the actual legal enactments are likely to differ considerably from the elegant theoretical models, if only for political reasons. (Compare the intellectually elegant *concept* of an income tax with the notorious complexity of the Internal Revenue Code!) The only way to see if these market solutions will work is to try them provisionally and carefully monitor the results.

Federalism is another possible form of decentralization. Today, most important standards are set in Washington. Federal regulations tend to be insensitive to differences in technological and economic constraints and to variations in environmental problems. More important, we also lose the chance to experiment with a variety of regulatory methods. The problem might be reduced by shifting more front-line regulatory authority to the states, subject to streamlined federal supervision.

The Clean Water Act illustrates one possible direction for reform. The current regulatory scheme gives the EPA control over pollution standards, with only limited discretion for state regulators. Under an alternate interpretation of the statute, the EPA could have set ranges of pollution for various industries as well as providing a list of factors to be used in making choices within that range. The states would then have chosen limits for individual plants within that range, subject to veto by the EPA. The Supreme Court rejected this alternative approach in 1977 partly because it was concerned about the impracticality of requiring the EPA to review thousands of state permits.[46] Today, however,

---

45. For arguments in favor of technology-based standards on these grounds, see Jerry Mashaw, "Imagining the Future; Remembering the Past," 1991 *Duke L.J.* 711, 721–723; Joel Mintz, "Economic Reform of Environmental Protection: A Brief Comment on a Recent Debate," 15 *Harv. Envtl. L. Rev.* 149 (1991); Sidney Shapiro and Thomas McGarity, "Not So Paradoxical: The Rationale for Technology-Based Regulation," 1991 *Duke L.J.* 729. For an appraisal of the flaws in existing programs and possible improvements, see Jeremy Hockenstein et al., "Crafting the Next Generation of Market-Based Environmental Tools," *Environment*, May 1997, at 13, 13.

46. See *E.I. du Pont de Nemours & Co. v. Train*, 430 U.S. 112, 132–133 (1977). Despite the limits of the state role under the current regulatory scheme, some

"[w]e can generate, analyze, and communicate a thousand times more information than we could just a generation ago, for a fraction of the cost."[47] Given modern methods of statistical quality control, the EPA may now have the effective capacity to oversee an individualized permit system.[48] The permit limitations established by different states could provide valuable new information, so that the EPA could adjust the guidelines based on experience.

Increased delegation could be used to help learn about possible new regulatory methods. The ideal of "states as laboratories" takes on a new relevance today. One thing we have learned about environmental regulation is that good ideas do not always work out. No matter how much we try to improve the regulatory process, many of our best ideas will fail, while less promising ideas sometimes will be unexpectedly successful. Or, more bluntly, we are always going to make a lot of mistakes. Given this reality, we ought to run a lot of experiments to test regulatory proposals.

There are obvious risks in delegating too much authority to states that may lack the resources, expertise, or political will to implement innovative environmental programs.[49] But these risks are not insurmountable. Subject to safeguards to prevent a "race to the bottom," we could give the EPA broad authority to con-

---

successful state innovation in water pollution control has taken place. See William Lowry, *The Dimensions of Federalism: State Governments and Pollution Control Policies* 73–78 (1992).

47. David Osborne and Ted Gaebler, *Reinventing Government: How the Entrepreneurial Spirit Is Transforming the Public Sector* 141 (1992).

48. Efficient systems to oversee even 42,000 permits (the number given by the *du Pont* Court, see 430 U.S. at 132–133) do not seem out of the question. Consider, for example, the vastly greater number of Medicare claims or income tax returns that must be screened annually. For example, the EPA might create a model to predict effluent limitations for plants having particular characteristics; the model could be based on economic or engineering theory, or it could incorporate statistical studies of actual permits from other states. Permits straying too far from the prediction would be automatically audited, as would a random sample of other permits. For a useful analysis of a similar proposal for the use of statistical claim profiles in tort cases, see Glen Robinson and Kenneth Abraham, "Collective Justice in Tort Law," 78 *Va. L. Rev.* 1481 (1992).

49. See Joshua Sarnoff, "The Continuing Imperative (But Only from a National Perspective) For Federal Environmental Protection," 7 *Duke Envtl. L. & Pol'y F.* 225 (1997).

tract with selected states to create innovative programs with carefully specified performance criteria.[50] States should be selected on the basis of demonstrated expertise and regulatory effectiveness. Successful state programs could then operate as models for other states or be incorporated into federal law. Unsuccessful state programs are nearly as important, since observing them may save us from making costly errors on a national scale.

The difficulties of designing these programs should not be underestimated. State governments differ widely in their capacity and will to engage in environmental regulation. Clear, enforceable performance standards are necessary to ensure that delegation does not turn into deregulation. In compensation for the risks attending delegation, we should probably insist that the results not only equal, but also exceed existing environmental targets. Environmental groups and representatives of local communities need to be heavily involved in the process to keep the "deals" honest and to ensure enforcement.

Arguments for decentralization are not uncommonly used as covert moves toward deregulation. By weakening federal regulators in favor of market mechanisms or state governments, some critics of the present system hope to weaken environmental regulation as a whole. That is far from being my purpose. I favor an increased degree of decentralization as a way of making environmental protection more effective, not as a way of undermining it. Making decentralization serve the cause of environmental protection will take careful design; it is far from being a simple matter of giving the states or the private sector a blank check.

Dynamic Regulation

No matter how much we try to decentralize, federal agencies like the EPA will still be making crucial regulatory decisions. The cur-

---

50. These contracts would be available only to states that had demonstrated the capacity to run an effective regulatory program. Such a contract would contain quantitative performance measures: specific levels of air or water quality to be met by particular dates. Failure to achieve these standards might result in penalties against the state or in cancellation of the contract. Finally, minimum federal standards would remain in place as a safeguard against risks to public health or irreparable environmental damage.

rent regulatory paradigm focuses on maximizing the quality of each individual agency decision. Except in a static situation, however, this may not result in the best regulatory outcomes over time. We need to move agencies toward a more dynamic mode, in which regulation is viewed as an ongoing cycle of experimentation and evaluation.

It seems rather painfully obvious that we cannot expect to improve environmental quality if we do not even know the *current* state of the environment. Unfortunately, our pollution monitoring is strikingly inadequate.[51] With regard to toxics, the EPA's information base was so weak that it was shocked when companies made the disclosures mandated by a new federal statute.[52] We urgently need better information about the present condition of the environment. Otherwise, we cannot measure the effectiveness of our current efforts, which is a necessary prerequisite of learning to do better.

We also need better follow-up on regulatory innovations. Market-based schemes present possibilities for making environmental protection more cost-effective without sacrificing environmental quality. But we badly need to know how well the schemes work in practice. Two leading environmental economists observed several years ago that, "[i]n spite of the potential importance of emissions trading as an alternative to conventional regulatory approaches, surprisingly little effort has been spent evaluating the impact of this program."[53] Program evaluation is not a luxury, but a necessity. Why continue to pour resources into programs with so little effort to evaluate their effectiveness? Researchers have begun to fill these gaps, but we should not have to rely on the efforts of individual researchers to

51. See Robert Percival et al., *Environmental Regulation: Law, Science, and Policy* 793, 866 (1992). For a survey of environmental monitoring, and calls for improvement, see Council on Environmental Quality, *Environmental Quality: Twenty-Second Annual Report* 43–56 (1992).

52. Percival et al., *supra* note 51, at 624.

53. Robert Hahn and Gordon Hester, "Where Did All the Markets Go? An Analysis of EPA's Emissions Trading Program," 6 *Yale J. on Reg.* 109, 109 (1989). For a similar, more recent assessment of the lack of program evaluation, see Office of Technology Assessment, *Environmental Policy Tools: A User's Guide* (OTA-Env-634, GPO 1995).

evaluate regulatory techniques. Systematic program evaluation should be built into every regulatory scheme, especially when innovative techniques are involved.

In an ideal world, the desirability of improved information would immediately translate into a higher budget for research and data collection.[54] Unfortunately, we seem to be moving in the wrong direction, as illustrated by the senseless congressional decision to close the Office of Technology Assessment.[55] Given budget realities, massive new support for environmental research is an unlikely prospect. Indeed, the EPA has had to struggle to return its research budget to pre-Reagan levels.[56] Some funds can be reallocated from other EPA activities, but this, too, has its limits. Consequently, we need to find ways to enlist industry in this process. Existing law contains several mechanisms for information generation. Pollution permits often require the monitoring and reporting of data, and these requirements could be expanded. (Indeed, there is precedent for requiring industry to finance research on new methods of pollution control.[57] Perhaps such a requirement could have been imposed on Reserve Mining.) Some other statutes also provide rough models for more creative requirements for data generation in the private sector.[58] Once information is obtained, we will need better systems for accessing the databases.[59]

54. On the ability of such improvements in information to increase social welfare, see Carlisle Ford Runge, "Economic Criteria and 'Net Social Risk' in the Analysis of Environmental Regulation," in *Environmental Policy under Reagan's Executive Order* 187, 198–201 (V. Kerry Smith ed., 1984).

55. See Robert L. Glicksman, "Regulatory Reform and (Breach of) the Contract with America," *Kan. J.L. & Pub. Pol'y*, Winter 1996, at 1, 11.

56. See Robin Shifrin, "Not by Risk Alone: Reforming EPA Research Priorities," 102 *Yale L.J.* 547, 563 n.72 (1992).

57. See *Kennecott Copper Corp. v. Train*, 526 F.2d 1149 (9th Cir. 1975), *cert. denied*, 425 U.S. 935 (1976) (upholding EPA requirement that firm undertake research program to improve pollution control technology).

58. See John Applegate, "The Perils of Unreasonable Risk: Information, Regulatory Policy, and Toxic Substances Control," 91 *Colum. L. Rev.* 261, 318–332 (1991).

59. See Mary Lyndon, "Information Economics and Chemical Toxicity: Designing Laws to Produce and Use Data," 87 *Mich. L. Rev.* 1795, 1840–1855 (1989). Environmental impact statements could provide another major source of environmental data if they were properly integrated into a unified database.

The possibility of acquiring relevant new information can also significantly change standards for decision making. Although the mathematical analysis is complex,[60] the basic idea is simple enough. If a decision would have irreparable consequences, then it may be worth waiting to obtain new information. Taking an irreversible step ends the possibility of future learning and therefore involves an extra cost that does not show up in the usual cost-benefit analysis. Waiting is equivalent to purchasing an option contract, and under many circumstances, that option is worth buying.

For example, suppose a cost-benefit analysis shows that a project has a 40 percent chance of producing a net million dollar loss and a 60 percent chance of a net million dollar gain. This looks like a good investment. A series of similar projects would produce average net gains of $200,000. On the other hand, suppose we can know the outcome of the investment with certainty if we wait six weeks. At that point, we can decide whether to go ahead or not. When we make the later investment decision, we will invest in the project 60 percent of the time, for an expected gain of $600,000, with no losses (since we will know enough not to invest in the loss situation). Waiting is the wise course here because making an immediate decision deprives us of the opportunity to obtain further important information.

The value of waiting can be dramatic. It is not unusual to find that an irreversible project should not be undertaken unless its expected benefits are at least twice its cost.[61] Otherwise, it is often better to wait for more information.

Given the magnitude of uncertainty and the likelihood of obtaining more information, the value of waiting may be quite important in environmental law. Destroying a rain forest or an

60. See Avinash Dixit, "Investment and Hysteresis," *J. Econ. Persp.*, Winter 1992, at 107, 107. The basic point is that "[w]here there is uncertainty, there may be learning." W. Kip Viscusi and Richard Zekhauser, "Environmental Policy Choice under Uncertainty," 3 *J. Envtl. Econ. & Mgmt.* 97, 108 (1976).

61. See Dixit, *supra* note 60, at 116. See also id. at 117, 120 for other examples of the magnitude of hysteresis effects. Sometimes we may be uncertain about the degree of irreversibility itself, and here, too, the possibility of learning must be taken into account. See Viscusi and Zekhauser, *supra* note 60, at 107–108.

endangered species is irreversible. Usually, whatever economic benefits can be obtained from exploiting the resource will be available if we wait, while the uncertainty about environmental costs will be reduced. Hence, there is a good argument for waiting while attempting to learn more.[62]

On the other hand, the value of waiting may disfavor certain forms of pollution control that involve large "sunk costs." Investments in pollution control equipment can't be recovered if it turns out that better technologies become available or that the harm caused by the pollution has been overestimated. We should try to avoid these stranded investments in environmental quality. Simply doing nothing while waiting for more information may be unacceptable, but we might consider less capital-intensive methods of control. Examples that come to mind include the use of respirators by workers to deal with airborne occupational hazards (rather than making massive investments in ventilation systems) and the use of low-sulfur coal rather than scrubbers to deal with acid rain. These stop-gap alternatives may not be the best solutions, but they can buy time while we seek more information.

In *Reserve Mining*, then, it might have made more sense to impose a more flexible remedy, providing for land disposal as an ultimate remedy unless in the meantime new information showed it was unfeasible or unnecessary. In the meantime, the court could have mandated temporary measures to reduce the risk. This strategy would have put pressure on Reserve to assist in exploring other remedies and in developing fuller information about the risk.

In *Corrosion Proof Fittings*, the EPA unsuccessfully tried to adopt a similar flexible strategy. One problem it faced was that the availability and safety of substitutes for some uses of asbestos were not fully known. Rather than delay all action until it had full information, it decided to issue a ban on these uses, but left the door open for variances if adequate substitutes proved un-

---

62. See Graciela Chichilnisky and Geoffrey Heal, "Global Environmental Risks," *J. Econ. Persp.*, Fall 1993, at 63, 76–79; Anthony Fisher and Michael Hanemann, "Option Value and the Extinction of Species," 4 *Advances in Applied Micro-Econ.* 169 (1986).

available. This technique put pressure on industry to develop and evaluate new substitutes, a task for which industry was much better suited than the agency. Unfortunately, the court seemed oblivious to the advantages of this regulatory technique, and reviewed the case as if the ban were ironclad, rather than a burden-shifting device.[63] Not surprisingly, the court found an insufficient basis for an absolute ban and struck down the asbestos regulations.

The Fifth Circuit's decision pushed the EPA in just the wrong direction. Rather than allowing the agency to take advantage of dynamic learning, the court forced it to invest its resources in making each individual regulatory decision perfect. It would have been more useful for the court to encourage the EPA to move toward even more flexibility, perhaps by encouraging a short-term freeze, rather than an immediate ban for some uses.

Corrosion Proof Fittings is not an isolated incident. Intensive judicial review has led to an ever-more elaborate, time-consuming regulatory process. Health standards that once could be issued in six months (from proposal to promulgation) now take six years instead.[64] It may seem obvious that an improved decision is always worthwhile. In a world of limited staff and budget, however, improvements in quality must result in delay and reduced output. Moreover, by the time all the data has been sifted and all the analytical bases have been covered, the world may have changed. As I argued earlier, sometimes waiting for more information is the wise course when the alternative is an irreversible change. But demanding perfection in decision making can also lead to paralysis when the situation allows a more experimental approach.

This lesson has been painfully learned in the private sector. Consider the rebuke given an auto executive who had helped delay his company's adoption of front-wheel-drive vehicles. "'The trouble with you,'" he was told, was that in business school "'they told you not to take any action until you've got all of the facts. You've got ninety-five percent of them, but it's going to take you another six months to get that last five percent. And

63. *Corrosion Proof Fittings v. EPA*, 947 F.2d 1201, 1220 (5th Cir. 1991).
64. Patricia Wald, "Regulation of Risk: Are Courts Part of the Solution or Most of the Problem?" 67 *So. Cal. L. Rev.* 621, 625 (1994).

by the time you do, your facts will be out of date because the market has moved on you.'"[65] Like the hapless target of this remark, the agency may find that the scientific data or the technological and economic constraints have shifted, leaving it with the choice of starting over or else closing the record and adopting a potentially obsolete regulation.

As in *Corrosion Proof Fittings*, agencies have been pushed toward static decision making by "hard look" judicial review, under which courts closely examine the administrative record to ensure agency rationality. Jerry Mashaw and David Harfst have detailed how the federal auto safety program was brought to a standstill by the judicial obsession with obtaining a full record.[66] It is not difficult to find cases like *Corrosion Proof Fittings*, in which the EPA's highly technical decisions have been overturned by courts demanding further documentation and more careful analysis.[67] Despite the initial appeal of hard-look review, there is substantial support for McGarity's recent appraisal that the "predictable result of stringent 'hard look' judicial review of complex rulemaking is ossification." Because agencies are never sure just how "hard" the hard look will be, they strive to protect themselves against the most unsympathetic judges. And "since the criteria for substantive judicial review are the same for re-

65. David Halberstam, *The Reckoning* 516 (1986) (quoting Lee Iacocca).

66. See generally Jerry Mashaw and David Harfst, *The Struggle for Auto Safety* (1990).

67. See, e.g., *Corrosion Proof Fittings v. EPA*, 947 F.2d 1201 (5th Cir. 1991) (overturning the EPA's carefully considered asbestos regulations, effectively wrecking its most serious effort to implement the Toxic Substances Control Act). It is probably not unfair to say that the Fifth Circuit's opinion "is so lacking in deference to the agency's exercise of expertise and policy judgment, and so full of attempts to impose on the agency the judges' own views of the proper role of regulation in society, that it is virtually indistinguishable from the documents that OMB prepares in connection with its oversight of EPA rulemaking." Thomas McGarity, "Some Thoughts on 'Deossifying' the Rulemaking Process," 41 *Duke L.J.* 1385, 1423 (1992). Other judges have not hesitated to correct agencies on technical issues like choice of the proper computer model. See, e.g., *AFL-CIO v. OSHA*, 965 F.2d 962 (11th Cir. 1992) (demanding that agency separately document health effects for each of 428 toxic substances, although OSHA argued that this was scientifically infeasible); *Ohio v. EPA*, 784 F.2d 224 (6th Cir. 1986) (rejecting EPA computer model); *Gulf S. Insulation v. Consumer Prods. Safety Comm'n*, 701 F.2d 1137 (5th Cir. 1983) (second-guessing the agency on technical issues).

pealing old rules as for promulgating new ones, agencies are equally chary of revisiting old rules, even in the name of flexibility."[68]

It is tempting to suggest abandoning the hard-look doctrine. The proper level of judicial review is, however, a complex question. Hard-look review has been used for various purposes, sometimes to keep the agency from ignoring its statutory mandate, sometimes (less persuasively) to improve the technical quality of agency decisions. Moreover, the proper scope of review is an issue that may arise in different settings, involving different agencies, statutes, and policy concerns. Despite its drawbacks, some form of hard-look review may sometimes be justified in some of these settings.[69]

A full exploration of the debate over judicial review would take us far beyond the scope of this book. In the context of environmental law, however, we may at least want to consider fine-tuning the level of review in order to foster the regulatory learning process. Rather than an across-the-board change in the level of judicial scrutiny, we might do better to vary the level of review depending on whether the agency is taking a dynamic or a static approach to regulation. A sensible approach would be to ease judicial review when an agency has a firm plan to monitor a rule's implementation and make appropriate modifications. Thus, we might lower the level of review when the agency shows the following:

- Its action will not cause irreparable injury.
- It has taken steps to generate additional relevant information.
- It has a process in place to reappraise current policy as the new information is developed.

Regulatory Reform and Agency Improvisation

One major problem in responding to new information is statutory rigidity. For example, most observers believe that Superfund

68. McGarity, *supra* note 67, at 1419–1420.

69. See generally Mark Seidenfeld, "Demystifying Deossification: Rethinking Recent Proposals to Modify Judicial Review of Notice and Comment Rulemaking," 75 *Tex. L. Rev.* 483 (1997).

is in need of serious revision.[70] By some accounts, the transaction costs are running between a quarter and a half of the direct cost of cleanup, and Peter Menell reports that given "rapidly escalating remediation costs . . . , the *transaction costs* of CERCLA's clean-up effort could exceed $44 billion."[71] This estimate may have been too alarmist, but no one questions that transaction costs have been substantial. Yet the amount of actual "cleaning up" to date seems to have been disappointing.[72]

Moreover, scientific knowledge has evolved since the statute was passed. The scientific premise of the statute was that toxic waste, and hazardous chemicals more generally, present an urgent public health risk. As we saw in the last section, however, scientific evidence has rapidly evolved, leaving environmental policy struggling to keep up. Although there are still grounds for concern, the risks seem less threatening now than they did in the crisis atmosphere when the statute was originally passed. One estimate is that the program costs between $384 million and $6.4 billion per life saved (depending on risk estimates and discount rates).[73] Focusing cleanups on the high-risk sites probably could cut costs dramatically with little effect on health.

Almost everyone agrees that the statute needs fixing to some degree, though debate remains about the extent of the necessary changes. But to date, Congress has not succeeded in revising the statute, despite the broad support for reform.[74] There is no question that the public continues to support the need for the cleanup program, and some of the practical implementation problems are now being ironed out administratively, but sub-

---

70. For an overview, see *Analyzing Superfund: Economics, Science, and Law* (Richard Revesz and Richard Stewart eds., 1995).

71. Peter Menell, "The Limitations of Legal Institutions for Addressing Environmental Risks," *J. Econ. Persp.*, Summer 1991, at 93, 108.

72. As of 1993, fewer than 70 of the 1,275 sites on the National Priorities List had been cleaned up. Rudy Abramson, "The Superfund Cleanup: Mired in Its Own Mess," *L.A. Times*, May 10, 1993, at 1A, col. 2. More recently the situation has been improving.

73. James Hamilton and W. Kip Viscusi, "The Benefits and Costs of Regulatory Reforms for Superfund," 16 *Stan. Envtl. L. Rev.* 159, 174 (1997).

74. See Rena Steinzor and Linda Greer, "In Defense of the Superfund Liability System: Matching the Diagnosis and Cure," 27 *Envtl. L. Rep.* (Envtl. L. Inst.) 10286 (1997).

stantial reforms are still needed, and Congress has seemed incapable of making them.

This kind of problem is a predictable side effect of our precautionary approach to environmental risks (an approach that I generally endorse). If we follow an environmentalist baseline and take a "conservative" approach to risk, we deliberately choose to err on the side of overprotection. Frequently, we will later discover that restrictions were too harsh. Unfortunately, such mistakes will often be made at the legislative level, where they are most difficult to correct.

Given what I said earlier in this chapter about prudent responses to uncertainty, I do not regard the system's bias toward overprotection as altogether regrettable. If err we must, we should err on the side of safety.[75] But even when it has later become clear that a statutory scheme is overprotective or too burdensome, Congress finds it very difficult to make the necessary correction. Recently, for example, it managed to enact partial reforms of two overprotective statutes, the Safe Drinking Water Act and the Delaney Clause (which banned all food additives capable of causing cancer at high doses in any animal). But the reforms took place only long after it was clear to the EPA itself and to informed observers that the earlier efforts were misguided and possibly even counterproductive.[76] Because environmental

75. The possibility of occasional public overreaction does not discredit the environmental baseline as a whole, partly because of the balancing counterinfluence of the regulated industries. The current baseline represents not merely the effects of initial public enthusiasm, but also the tempering experience of implementing regulatory schemes. Moreover, in formulating an environmental baseline, the main point is not whether particular statutes are scientifically sound, but rather what price the public is willing to pay to achieve perceived environmental gains. In short, the baseline that I advocate has more to do with the public's values than with the accuracy of its factual assessments. Those public value judgments mandate the kind of protective attitude toward the environment advocated in chapter 4. But embrace of environmentalist goals does not mean that every particular statutory provision is initially well considered, let alone that every provision will stand the tests of time and developing knowledge.

76. See Scott Bauer, "The Food Quality Protection Act of 1996: Replacing Old Impracticalities with New Uncertainties in Pesticide Regulation," 75 *N.C. L. Rev.* 1369 (1997); Andrew Miller, "The Food Quality Protection Act of 1996: Science and Law at a Crossroads," 7 *Duke Envtl. L. & Pol'y F.* 393 (1997); A. Dan Tarlock, "Safe Drinking Water: A Federalism Perspective," 21 *Wm. & Mary Envtl. L. & Pol'y Rev.* 233 (1997).

statutes are increasingly complex and detailed, this problem is even more likely to occur at the micro-level. Among the host of specific mandates imposed by Congress in an effort to restrain administrative discretion are some that misfire.

As a form of deregulation, the Contract with America called for rigorous cost-benefit risk analysis as a "supermandate" for existing regulations. As I have already made clear, even for new regulations, this supermandate was misguided. It badly overestimated the existing state of the art for cost-benefit analysis, overlooked the importance of hard-to-quantify values in environmental regulation, and shifted too much decision making away from front-line regulators to OMB economists and reviewing judges. As a method for reforming with *existing* regulations, it was even worse. It would simply have clogged the agency's docket with impossible demands for intensive review of past efforts.

Another approach is the use of sunset provisions to trigger additional congressional oversight. This sounds good in theory, but hasn't worked well in practice. Under this approach, program financing must be renewed on a periodic basis. Unfortunately, Congress hasn't succeeded in using these renewals constructively. Superfund, for example, has been up for renewal for years, but limps along under the current statute because Congress has deadlocked over changes.

Our unwieldy political process cannot easily keep up with rapidly changing knowledge about environmental problems and how to address them. It may be especially hard for Congress to back down from excessive regulation, given the fear of being accused of caving to anti-environmentalist special interests. Policy changes are especially difficult for Congress to achieve in an era of divided government. The words "nimble" and "legislature" simply do not fit comfortably together. And yet in such a rapidly changing field, we need a nimble response.

Congress *could* achieve flexibility by giving agencies like the EPA broader discretion, leaving it to the agencies to make the basic policy decisions. Although this would increase flexibility, it may be politically unacceptable. It also would dissipate the important idea of an environmentalist baseline, which was painstakingly established over decades in a host of environmental statutes. Rather than eliminating this environmentalist baseline, we need a mechanism by which the EPA could modify its choice of means, but only where changed circumstances clearly justify a modification.

A model for such a mechanism exists in telecommunications law, another area where it is very difficult for Congress to keep up with rapid change. Before 1996, the industry was governed by the Communications Act of 1934, which treated long-distance carriers like public utilities. This made sense when AT&T had a virtual monopoly, but not after the advent of serious competition. In an attempt to deregulate, the Federal Communications Commission (FCC) invoked its "modification" power under the 1934 law, which authorized it to "modify any requirements made by or under" the rate regulation provision. The FCC argued that deregulation was a "modification" of the statute's requirements. In one of Justice Anton Scalia's typical exercises in literalism in statutory interpretation, he consulted several dictionaries, found only one in which the word "modify" was used this broadly, and struck down the agency's interpretation.[77] In dissent, Justice John Paul Stevens protested that the majority's "rigid literalism" deprived the FCC of "the flexibility Congress meant it to have in order to implement the core policies of the Act in rapidly changing conditions."[78]

Congress apparently agreed with Justice Stevens. In 1996, it passed a massive new telecommunications act. Title IV of the statute is entitled "Regulatory Reform." The crucial language authorizes the FCC to exempt carriers from a regulation or statutory requirement if three conditions are met:

(1) enforcement of such regulation or provision is not necessary to ensure that the charges, practices, classifications, or regulations by, for, or in connection with that telecommunications carrier or telecommunications service are just and reasonable and are not unjustly or unreasonably discriminatory;
(2) enforcement of such regulation or provision is not necessary for the protection of consumers; and

---

77. *MCI Telecommunications v. AT&T,* 512 U.S. 218, 225–228 (1994).
78. Id. at 235 (Stevens, J., dissenting).

(3) forbearance from applying such provision or regulation is consistent with the public interest.

In making this determination, the FCC must consider whether its action will promote competitive market conditions.[79]

Congress should give the EPA similar powers to engage in regulatory reform. The standard for the EPA's action should be a showing that applying the existing regulatory scheme is (1) unnecessary because it has become clear that no significant threat exists to the environment or public health or (2) infeasible because its burdens have turned out to be grossly disproportionate to any possible environmental problem and a less burdensome alternative exists. In other words, the EPA would deregulate when a federal scheme as a whole no longer implemented the environmental baseline of feasibly regulating all significant risks.

This power could be used for various purposes. The EPA could waive existing regulations after entering into special contracts with companies to use other forms of pollution control, as in the Clinton administration's Project XL.[80] One example is the EPA's innovative agreement with Intel. The chipmaker agreed to strict environmental goals in return for increased flexibility in choosing how to meet the goals.[81] Although the EPA has experimented with such contracts, it has been handicapped by lack of statutory authorization.[82] Moving beyond agreements with individual polluters, the EPA could also approve state programs that were more cost-effective than the federal scheme. In the long run, the thrust of the EPA's actions could conceivably switch

79. 47 U.S.C.A. § 160 (West 1998). For discussion of the statute and its background, see Jim Chen, "The Legal Process and Political Economy of Telecommunications Reform," 97 *Colum. L. Rev.* 835 (1997).

80. See Rena I. Steinzor, "Regulatory Reinvention and Project XL: Does the Emperor Have Any Clothes?" 26 *Envtl. L. Rep.* (Envtl. L. Inst.) 10527 (Oct. 1996); Joel A. Mintz, "Rebuttal: EPA Enforcement and the Challenge of Change," 26 *Envtl. L. Rep.* (Envtl. L. Inst.) 10538 (Oct. 1996). For a thoughtful appraisal of Project XL and related efforts, see Jody Freeman, "Collaborative Governance and the Administrative State," 45 *UCLA L. Rev.* 1 (1997).

81. See John Cushman, Jr., "EPA Innovates at Big Arizona Factory," *N.Y. Times,* Nov. 20, 1996, at A10.

82. See Bradford Mank, "The Environmental Protection Agency's Project XL and Other Regulatory Reform Initiatives: The Need for Legislative Authorization," 25 *Ecology L.Q.* 1, 5 (1998).

from enforcing regulations to policing risk-reduction arrangements created by industry and state governments.[83]

As with decentralization, my motivation for providing the EPA with greater flexibility is not to undermine environmental protection, but to strengthen it. The EPA's ability to protect the environment is hampered by the rigidity of today's highly complex regulatory statutes, especially given the increasingly inflexible and literalistic approach to interpretation followed by the courts these days. When experience shows that a given statutory mechanism is unworkable, the EPA should be able to search for alternative means to achieve the congressional goal. The EPA should also be able to shift resources to more valuable programs when a particular program's environmental benefits prove illusory. In the long run, assuming the provision can be structured to minimize the risks of abuse, it should lead to a stronger overall program of environmental protection.

This provision would grant a significant new power to the EPA, and this degree of discretion might be considered dangerous. There are two responses to this concern. First, the EPA has proved itself to be a reasonably reliable trustee for the environment. The main exception was the first couple of years of the Reagan administration. Even then, political forces proved capable of reining in the worst abuses. Second, the EPA should have to make a convincing showing in order to invoke this provision, since it would be cutting against the dominant baseline of environmental law. Judicial review should provide a substantial safeguard against abuse of this power, given the EPA's burden of justifying its action.[84]

Neither safeguard against abuse is foolproof. The political situation has changed since the Reagan administration, and today's Congress might be less interested in reining in administrative abuse. On the other hand, politicians across the political spec-

83. See E. Donald Elliott, "Toward Ecological Law and Policy," in Marion Chertow and Daniel Esty, *Thinking Ecologically: The Next Generation of Environmental Policy* 170, 180–181 (1997).

84. Admittedly, hard-look judicial review may hinder some desirable regulatory reforms. As we gain experience with deregulation, more nuanced scrutiny by the courts may evolve. Initially, however, the promise of judicial oversight should help allay fears about deregulation.

trum have come to appreciate the strength of the public's attach-ment to environmental values, partly as a result of the Reagan experience. Thus, future EPA administrators may be less in-clined to play fast and loose with the environment. Courts pro-vide another line of defense, but doctrines of administrative law give agencies some leeway to pursue their own goals. Moreover, judges differ in their abilities and their attitudes toward environ-mental protection. Thus, the risks posed by increased EPA flex-ibility cannot be entirely discounted. Nevertheless, the risks of regulatory ossification in a fast-changing world do need to be ad-dressed.

To the extent readers remain worried that this is too much power to place in the EPA, I would suggest that the concern re-ally relates more to the EPA generally than to this specific pro-posal. The EPA already has an enormous amount of discretion, and if we cannot trust it to exercise such discretion respons-ibly, we have a problem that goes well beyond the question of the EPA's involvement in regulatory reform. Perhaps, if these concerns are justified, we should consider some institutional changes, such as assigning environmental protection to an inde-pendent agency, rather than one run by a political appointee. With respect to the proposal for regulatory flexibility more spe-cifically, we could also add various procedural safeguards such as special hearing requirements in order to lower the chances of abuse. But my own sense is that the political actors have learned by now that the public responds angrily to efforts to undermine the environment and that the risk of EPA abuse of the power to engage in regulatory reform is limited enough to be worth taking.

Giving the EPA authority for regulatory reform would address the fear that the existing regulatory system has become too cum-bersome, expensive, and overprotective. Yet the EPA would con-tinue to act against the background of a congressionally man-dated environmentalist baseline. The existence of an escape valve might even strengthen support for the environmental base-line by making it clear that later adjustments would be available.

We have been concerned so far with situations in which regu-lations turn out to be unexpectedly burdensome or unnecessarily strict. What about unanticipated environmental problems (including problems that turn out to be surprisingly severe)? Ultimately, we can expect Congress to take action, if past experience is any guide, but the EPA may need to take some steps in the meantime. A number of environmental statutes contain "imminent hazard" provisions, allowing the EPA to obtain injunctions against dangerous conduct. The *Reserve Mining* suit was brought in part under one of these provisions. Conceivably, an unanticipated environmental threat might not fall within any of these existing statutes. There is some argument for consolidating and broadening the existing statutes, with a law covering any imminent environmental risk. Tort law also exists as a backstop to the regulatory system. As *Reserve Mining* illustrates, however, courts are likely to find the issues raised by these suits knotty, and this approach should be considered only a stopgap until better regulatory techniques can be put in place to deal with the problem.

As we have seen, one of the great challenges of environmental law is the pervasive uncertainty that surrounds environmental issues. Sometimes all we can do is to make the best decision we can with whatever information we have at hand. When hard numbers are unavailable, this may mean taking reasonable safeguards against serious, but unquantifiable, risks, perhaps placing the burden of proof on the polluter. When we do have better information, we can use the environmental baseline of chapter 4, taking all feasible measures against significant risks unless the costs are clearly disproportionate to possible benefits.

But in the long run, the structure of decision making may be more important than what test is applied in individual cases. We need to create structures of decision making that allow us to take advantage of increased knowledge over time. Such structures may include various forms of decentralization, administrative strategies that allow options to be kept open, and deregulatory authority to eliminate outmoded regulatory requirements. These types of flexibility should not be seen as hostile to environmentalism. Instead, they are ways of maintaining the vitality of environmental protection. Rather than being a strength, excessive rigidity could lead to collapse under the pressure of constant change. A more flexible regulatory system, in the long run, may provide a higher level of environmental quality.

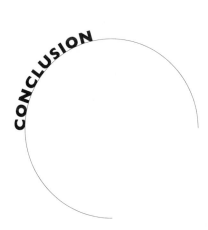

CONCLUSION

     The premise of this book is that environmental law is here to stay. Admittedly, prediction is a hazardous enterprise. Maybe environmental law will vanish in some political conflagration. So far, however, it has shown strong staying power, having survived the Reagan Revolution (as well as the "Gingrich Aftershock") virtually unscathed. It would be tragic if we did lose our national commitment to the environment, but currently this outcome seems unlikely. It is now time to consider how to shape our regulatory system to implement this commitment most effectively for the indefinite future.

    In mapping out the future of environmental law, we need to consider not only how to make the best environmental decisions at any given time, but also how to create a sustainable environmental law that can endure over the long haul. To be sustainable, environmental law must accommodate not just environmentalism, but also other key values. Otherwise, it will slowly erode.

    In this book, I have advocated a pragmatic approach to environmental regulation, but one grounded in environmentalism. The key norm is that we are all presumptively entitled to a safe environment and to the preservation of nature. This norm is now firmly embedded in our political culture. But the norm is tempered by an awareness of competing goals.

    As a society, we have complex and sometimes conflicting views about environmental issues. For the past thirty years, we have collectively expressed a fervent belief in preserving nature and protecting human health. We find it unjust for some people to reap profits by threatening the health of others through pollution or by destroying wilderness and seacoasts that belong to everyone. These views are not merely a passing political fad. They seem to be deeply rooted in the national psyche. These views may connect with the historical role of wilderness in American culture, and also perhaps with an instinctive human attachment to certain kinds of animals and landscapes. They reflect the strong American belief in personal autonomy and physical integrity, which are threatened by involuntary exposure to pollution. In this book, I have not felt it necessary to argue at length for the legitimacy of this national commitment, but it underlies most of the positions I have taken.[1] For this reason, I have called for analyzing regulatory issues from an environmental baseline.

At the same time, however, there are limits to how far Americans are willing to go to achieve environmental goals. Our political system avoids imposing severe losses on particular economic sectors, whether in the form of corporate bankruptcies or individual layoffs. As discussed in chapter 3, our society does have a strong commitment to environmental values, but it also recognizes the legitimacy of economic goals. For better or worse, there is no sign that we are prepared to give up what American consumers now regard as the basics of the good life. Economic growth is not something we are prepared to abandon in the name of environmental protection. We believe in the future, but we also want to live well in the present. For these reasons, as I argued in chapter 3, environmental protection cannot take an absolutist stance without departing from the mainstream of American political culture.[2]

1. For readers seeking a discussion of the arguments for environmental preservation, good starting points include Christopher Stone, *Earth and Other Ethics: The Case for Moral Pluralism* (1987); Andrew Brennan, *Thinking about Nature: An Investigation of Nature, Value, and Ecology* (1988); David Takacs, *The Idea of Biodiversity: Philosophies of Paradise* (1996).

2. Although rarely found in the legal academy, there are more radical versions of environmentalism that do take such an absolutist stance. For a good description and critique of "deep ecology" and "eco-feminism," see Martin Lewis, *Green Delusions: An Environmentalist Critique of Radical Environmentalism* (1992).

Thus, there is a pervasive tension in our responses to environmental problems. This tension is not just expressed as a conflict between different political groups. It is also internal to the personal views of most of us. In *Reserve Mining*, it is hard not to be repelled by the company's massive dumping in the pristine waters of Lake Superior, as if this great natural wonder were merely its private garbage pit. Yet the company's claims have their own appeal: it began its activities with the full approval of society, it was using only a tiny corner of a very large lake, the materials were (at least superficially) innocuous sand and gravel, and the evidence of any harm to health or ecology was unclear. Perhaps these considerations sway us in favor of the company. But yet again, we cannot be sure that Reserve's activities were harmless, and there was always the outside chance of some catastrophic outcome. But then still again, isn't some level of risk an inevitable part of life? So we are tugged first one way and then the other. As a society, we currently seem to be pulled more forcefully by the environmentalist side of this tug-of-war, but pulled both ways nonetheless.

If environmental law is to do justice to our society's complex views, it must also reflect this tension between environmentalism and economics. We must be wary of too facile a resolution of the conflict. But we do need some "rules of engagement" for environmental regulators. Thus, environmental pragmatism needs to keep us firmly aware of the complexity of our values, but prevent that complexity from entangling us to the point of inaction. Over time, the rules of engagement will surely shift as society's norms continue to evolve. No set of rules can be expected to make hard problems suddenly easy. But what we can seek are principles to guide our judgment in particular cases.

In more concrete terms, eco-pragmatism translates into several guidelines for environmental policy:

- When a reasonably ascertainable risk reaches a significant level, take all feasible steps to abate it except when costs would clearly overwhelm any potential benefits. Meanwhile, take prudent precautions against uncharted, but potentially serious, risks.
- Take a long-range view. Use low discount rates, maintain the responsibility of the current generation to ensure a

liveable future, and treat the preservation of nature as an opportunity for long-term social saving.

• Keep in mind the uncertainty surrounding many environmental problems. Adopt coping strategies such as burden-shifting rules, postponement of irreversible decisions, and (when appropriate because of new information) deregulation.

• Overall, keep a sense of balance, while maintaining a firm commitment to environmentalism. Don't put economists in charge of the regulatory process, but take their views seriously as a reality check on overzealous regulation.

This is an environmentalist program, but in a "pragmatic shade of green." It recognizes the need to temper environmental protection in the name of other values. It includes precautionary strategies, but it also provides for regulatory reform when new information shows those precautions were unneeded or poorly designed. Willingness to correct overregulation is an important element of the program—without it, environmental law would inevitably be weighed down by obsolete strategies that would not only overburden regulators, but also produce a political backlash. A viable environmental program needs to contain its own corrective mechanisms.

The pragmatist side of this blend is not simply a result of my general philosophical leanings. It is also a response to the need to make environmental law durable enough to function over decades and longer. The need for regulatory durability supported the arguments for giving weight to economic interests in chapters 2 and 4, for restraining demands for intergenerational sacrifices in chapter 5, and for proposing to increase regulatory flexibility in chapter 6. We simply cannot afford to lose sight of the need for environmental protection to be sustained over time.

It may be appropriate, in closing, to discuss some of the broader implications of the idea of sustainable environmental regulation. We are in the business of environmental protection for the long haul. Threats to biodiversity, to air and water quality, and to human health are not about to evaporate. Population pressures and economic growth will probably intensify the threats. Fundamentally, environmental protection will come to very little unless it can be entrenched. What good does it do to preserve an endangered species today unless we can make a continued commitment to protecting it in the future? The shadow of the future looms large in environmental law.

In discussing intergenerational duties in chapter 5, I stressed the need to scale our commitments so we can realistically expect them to retain popular support. We should not expect extraordinary sacrifices to benefit future generations, but we should insist on maintaining a decent minimum for those who follow us. Like long-distance runners, we need to budget our effort for a marathon, not a quick sprint.

Similarly, if we try to achieve a zero level of risk, we are likely to find the effort flagging in a few years. Heroic sacrifices can be expected in emergencies (and perhaps on a daily basis from saints), but not for entire populations over long periods of time. So we need to leave room for cost-benefit analysis as a restraint on overzealous regulation, without allowing it to become a straitjacket. Likewise, if we lock ourselves into outmoded legislation based on stale science, we risk discrediting the entire enterprise over the long term.

To be sustainable, environmental regulation cannot rely too heavily on outright coercion. Over long periods of time, it is too difficult to maintain a strict enforcement regime against strong opposition by the regulated community. If we want to protect endangered species, we need to enlist the support of the people who live with them, whether in the American West or the Amazon basin. This means finding ways to make preservation fit with their interests and goals, rather than making it a hated foreign intrusion. It also means continued attention to the fairness of the regulatory requirements themselves and of the enforcement process.

To be sustainable, environmental law also needs to maintain its credibility with the public. The biggest challenge may be the "democracy deficit" in environmental law. In their broad outlines, environmental statutes may represent popular sentiment, but the details are appallingly complex, and the technical issues are remarkably tough. We cannot expect the average citizen to understand the byzantine details, and yet we cannot afford to leave people alienated from a technocratic control regime. We can and should simplify regulation as much as possible in the interests of comprehensibility.

We can also promote public education and try to involve

people as much as possible in decisions affecting them on the local level. Education used to be offered as a kind of liberal panacea to all social problems. We know now from experience that this remedy has its limits. Nevertheless, we should not lose sight of the need to educate the public about environmental problems and their potential solutions. This educational effort should go beyond the "save the planet" slogans to encourage critical thinking about environmental issues.[3] We can also help inform the public through broadening disclosure requirements, implementing public hearings, and using other techniques to expand participation in environmental decision making.[4]

Another strategy involves reinforcing the credibility of environmental agencies themselves. This means stronger support for research and peer review and more oversight by outside evaluators like the National Academy of Science. It also means protecting the EPA from political interference by strengthening its sources of funding and preventing a repeat of the crass politicization of the early Reagan years. For example, as I mentioned in chapter 6, we might think about making the EPA an independent commission, rather than an agency run by a political, cabinet-level appointee. We need to give people good grounds for trust in environmental regulations.

We also need to promote the vitality of environmental values. Many of us are "armchair environmentalists," rather than wilderness backpackers. Indeed, society cannot afford too many backpackers without putting the wilderness itself at risk. Yet people

3. For a discussion of the need for such education and of efforts in that direction, see Willett Kempton, "How the Public Views Climate Change," *Environment*, Nov. 1997, at 13.

4. There are limits, however, to how much information we can expect people to master. Hence, we should also pursue other possible strategies to maintain public credibility. One is to reinforce the role of environmental groups as monitors of the government. People may be unable to judge whether a new pollution regulation will work, but they may be willing to accept the word of the Sierra Club that it's a useful improvement. If environmental groups are to play this role effectively, however, they need sufficient access to information themselves to make intelligent judgments. They also need to be weaned away from the need to publicize the "crisis of the month" to garner public support. These considerations argue for giving them more of a role in the regulatory process, including some financial support.

cannot be expected to support environmental protection unless nature is somehow part of their lives. Perhaps we should promote Internet contact with the environment as well, through real-time virtual viewings of wilderness. In an increasingly crowded world, not only is it an increasing challenge to preserve nature, but also it may be even more of a challenge to preserve opportunities to experience nature without overloading it.

Environmental purists may not take much interest in zoos or city parks, but these settings do provide important opportunities for contact with some aspects of nature, as do nature shows in the media. We don't normally think of environmentalism as having much to do with city parks because we are used to imagining a hermetic barrier: on the one side, unsullied nature, preserving its delicate balance, and on the other, the sordid world of human activity. But the current teaching of ecologists is that this picture is all wrong. Nature is not in an equilibrium; it is in a constant state of change. The old picture of the "balance of nature" simply does not correspond to reality. Nor is nature unsullied by humans. Humans have appropriated over a third of the earth's surface and now command about the same share of its biological production.[5] Except on Antarctica, there is no ecosystem that has not been profoundly shaped by the human presence—not just by modern technology, but also by the hunting and fires of aboriginal peoples. The North American forests that the Pilgrims found were a product of fires and other activities of the Natives Americans, not a primeval remnant of prehuman times.[6]

There are two lessons here: we need to think of human society as firmly embedded in nature, and we need to think of nature as a flux rather than a balance. So environmentalism cannot take the form of a "Berlin wall" keeping the humans out and the animals in. Instead, we must envision long-term connections between humans and nature, requiring continual change and adaptation on both sides. If the environment is to be sustainable,

5. Peter Vitousete et al., "Human Domination of the Earth's Ecosystems," 277 *Science* 494 (1997).

6. For discussions of the "New Ecology," see the "Symposium on Ecology and the Law," 69 *Chi.-Kent L. Rev.* 847–985 (1994); Daniel Botkin, *Discordant Harmonies: A New Ecology for the Twenty-First Century* (1990); Andrew Dobson, *Conservation and Biodiversity* (1996).

we will need institutions that are also sustainable: firmly entrenched, yet flexible and capable of adaptation.[7]

At its core, eco-pragmatism is an effort to integrate human society and the environment in a way that will benefit both. Nature and human society must be seen as interpenetrating, so that both are shaped by our own values as well as by natural processes. This necessarily requires a process of mutual adaptation. Human activities and values must be suitable for sustaining the planet, but the state of the planet must also be suitable for sustaining modern human societies.

The most fundamental key to environmental sustainability is having a set of values that we can expect people to live by. Cost-benefit analysis does not fit the bill. The number of people who will ever be willing to plot their lives to maximize economic efficiency is, thankfully, quite limited. But a fanatic devotion to the environment is not only an unrealistic expectation, but also an unhealthy one. Pragmatism affirms the complexity of our values, which encompass present consumption, future welfare, and the intrinsic worth of nature. This is a rich enough vision to provide a permanent grounding for environmental preservation.

Difficult trade-offs must be made, trade-offs that are made even more difficult by massive scientific uncertainty. And yet if we can make these trade-offs intelligently, remaining faithful to our commitments, but with one eye on the future, we may find value in the very act of decision itself. Wise decisions are not easy, but to decide wisely is of value in and of itself. To cope with environmental challenges, we will need a society that is attached to environmental norms, willing to take a long-term perspective, and institutionally capable of wise decisions. This ideal sometimes seems far removed from reality. It is easy to be cynical about politics. But we have made considerable progress in addressing environmental problems. Although being pragmatic means being realistic, it does not mean abandoning our hopes for the future.

7. See A. Dan Tarlock, "Environmental Law: Ethics or Science?" 7 *Duke Envtl. L. & Pol'y F.* 193 (1996).

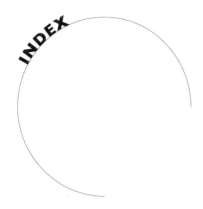

INDEX